D1307454

New View: Hypercomplex Cosmology Based on the Unification of General Relativity and Quantum Theory

Stephen Blaha Ph. D.
Blaha Research

Pingree-Hill Publishing
MMXXII

Rev. 00/00/01   January 18, 2022

To Margaret

Some Other Books by Stephen Blaha

All the Megaverse! Starships Exploring the Endless Universes of the Cosmos using the Baryonic Force (Blaha Research, Auburn, NH, 2014)

SuperCivilizations: Civilizations as Superorganisms (McMann-Fisher Publishing, Auburn, NH, 2010)

All the Universe! Faster Than Light Tachyon Quark Starships & Particle Accelerators with the LHC as a Prototype Starship Drive Scientific Edition (Pingree-Hill Publishing, Auburn, NH, 2011).

Unification of God Theory and Unified SuperStandard Model THIRD EDITION (Pingree Hill Publishing, Auburn, NH, 2018).

The Exact QED Calculation of the Fine Structure Constant Implies ALL 4D Universes have the Same Physics/Life Prospects (Pingree Hill Publishing, Auburn, NH, 2019).

Unified SuperStandard Theory and the SuperUniverse Model: The Foundation of Science (Pingree Hill Publishing, Auburn, NH, 2018).

Quaternion Unified SuperStandard Theory (The QUeST) and Megaverse Octonion SuperStandard Theory (MOST) (Pingree Hill Publishing, Auburn, NH, 2020).

Unified SuperStandard Theories for Quaternion Universes & The Octonion Megaverse (Pingree Hill Publishing, Auburn, NH, 2020).

The Essence of Eternity: Quaternion & Octonion SuperStandard Theories (Pingree Hill Publishing, Auburn, NH, 2020).

From Octonion Cosmology to the Unified SuperStandard Theory of Particles (Pingree Hill Publishing, Auburn, NH, 2020).

Beyond Octonion Cosmology (Pingree Hill Publishing, Auburn, NH, 2021).

Integration of General Relativity and Quantum Theory: Octonion Cosmology, GiFT, Creation/Annihilation Spaces CASe, Reduction of Spaces to a Few Fermions and Symmetries in Fundamental Frames (Pingree Hill Publishing, Auburn, NH, 2021).

Available on Amazon.com, bn.com Amazon.co.uk and other international web sites as well as at better bookstores (through Ingram Distributors).

CONTENTS

FIGURES and TABLES

Introduction

The author introduced the first hypercomplex number based theory for symmetries and particles in January, 2020. Since then he developed a comprehensive theory described in a number of books that became Octonion Cosmology. Octonion Cosmology was based on spinor spaces. In the past few months the author has developed a new theory: Hypercomplex Cosmology (HC) based on the CASe groups within GiFT that are associated with spaces of creation/annihilation operators in PseudoQuantum Field Theory.

HC has a much better set of 10 spaces and a more transparent derivation of their properties. It accounts for the known form of the Standard Model exactly without any *ad hoc* choice of symmetries. It generalizes the Standard Model to the Unified SuperStandard Theory (UST) theory with 256 fundamental fermions and a large set of Internal Symmetries in a 256 dimension space containing their irreducible representations. HC generates universes in "sister" pairs through fermion-antifermion annihilation.

Quantum theory is based on quantum fields, which contain creation/annihilation operators. These operators undergo General Relativistic transformations that generate spaces containing particles and symmetries. The dynamics of the particles and symmetries are governed by a Riemann-Christoffel curvature tensor. General Relativity appears in the dynamics. The unification of Quantum Theory and General Relativity is completed by the combination of space-time dimensions and internal symmetry representation dimensions in a dimension array in each HC space..

HC also supports a simpler formulation of quantum scalar and fermion space fields.

HC determines the exact known particles and symmetries of the Standard Model and adds more to form a complete unified theory. In particular, it uses CASe su(1, 1) symmetry to define a two-time coordinate system for a Fundamental Reference Frame. This coordinate system directly implies the Standard Model symmetries and the form of their fermion spectrum.

Among other advantages, HC General Relativistic transformations map the known fermion and symmetry structure to a much reduced number of fermions and symmetries – a one generation Standard Model in a non-static reference frame called the Fundamental Frame by the author. The number of space-time dimensions in a space, the total number of internal symmetry dimensions of a space, and the number of fundamental fermions of the space's Fundamental Frame are linked thus supporting the reality of the Fundamental Frame.

Hypercomplex Cosmology completely accounts for the fermion spectrum and symmetries of the Unified SuperStandard Theory (and QUeST.) HC unifies all the known interactions including Gravitation.

The ten Hypercomplex Cosmology spaces depicted on the cover remind the author of the Bohr Theory hydrogen energy levels that preceded Quantum Mechanics. The author believes there is a deeper theory beneath HC that remains to be found.

Appendix A contains excerpts from *Integration of General Relativity and Quantum Theory: Octonion Cosmology, GiFT, Creation/Annihilation Spaces CASe, Reduction of Spaces to a Few Fermions and Symmetries in Fundamental Frames*. Appendix B contains references to the author's original papers on PseudoQuantum Theory of the 1970's. Appendix C describes features of PseudoQuantum Theory and Two-Tier Theory.

1. Unification of Quantum Theory and General Relativity

This work unifies Quantum Theory and General Relativity within the GiFT formulation. It presents a new view: Hypercomplex Cosmology (HC) and its spectrum of spaces.

In prior books the spaces of Octonion Cosmology were developed based on the correspondence between fermion spinor arrays and spaces based on hypercomplex number arrays. We showed that spinor arrays could be views as arrays of hypercomplex numbers.

In this book we develop a new view based on the CASe spaces of the GiFT formulation of Quantum theory and General Relativity. In this view Hypercomplex Cosmology spaces are generated from CASe group transformations. CASe group transformations embody implicit General Relativistic transformations among static and non-static coordinates systems (reference frames). CASe transformations transform Creation and Annihilation operators, which implement Quantum theory at the most basic level. Thus the unification within GiFT.

At the deepest level the fundamental requirements for a theory of Physics are:

1. A particle formulation of matter and energy that leads to Quantum Theory.
2. A set of coordinate systems that embody General Relativity to ensure that physical processes are equivalent in the various coordinate systems: General Covariance.

As we have discussed previous particle creation/annihilation operators are embodied in quantum field expansions; quantum fields require Quantum Field Theory; and Quantum Field Theory is the basis of *physical* Quantum Mechanics.[1]

The formulation of Quantum Field Theory within the GiFT PseudoQuantum framework is needed for a number of important reasons.[2]

An additional reason, which is evident in Blaha (2021j) and Appendix A here, is it provides a formulation of CASe spaces that dovetails exactly with the Hypercomplex Cosmology spectrum of spaces with a superficial but important change. In Fig. 1.1 we display the new view of spaces. Note that the previous anomaly of space-time dimension in Octonion Cosmology (that required compacted dimensions) is now eliminated. Fig. 1.1 should be compared to Fig. 4.1 in Appendix A, which first appeared in Blaha (2021j).

In each space of Hypercomplex Cosmology a fundamental fermion has a set of creation and annihilation operators that undergo transformations when a General

---

[1] Quantum Mechanics can be studied as an independent mathematical theory. Any ambiguities/issues that arise in that study must be resolvable physically since Nature has only one physical "answer" in any physical phenomena.

[2] Section C.11.3 of Appendix C lists some of the reasons. The PseudoQuantum Theory is outlined in Appendix C.

Relativistic transformation between a static and a non-static reference frame is made. The set of such transformations has a su($2^{r/2}$, $2^{r/2}$) symmetry group in the space with r space-time coordinates. The irreducible representation of the symmetry has $2^{r/2 + 2}$ real dimensions.

The result is the NEWQUEST and NEWUST theories presented in our earlier books. Also the Megaverse theory NEWUTMOST follows for Cayley-Dickson number 4 space with the change in space-time dimensions from 8 to 6. (The Connection Groups for this space are also modified.) Thus we obtain the theories presented earlier with minor changes but based on a different principle.

1.1 Unification

The unification of General Relativity and Quantum Theory is evident in Hypercomplex Cosmology at several levels. The complete set of symmetry groups, including the space-time group, is contained in each space. The dimension array with $2^{r/2 + 2} \times 2^{r/2 + 2} = 2^{r+4}$ real dimensions is the irreducible representation of a U($2^{r + 3}$) symmetry group that is broken into the set of subgroups for internal symmetries and space-time for each Hypercomplex Cosmology space. We view the U($2^{r + 3}$) group as the initial group at the point of creation – the Big Bang, which has extremely large energy concentrated at one point.[3]

The explicit unity of the symmetry groups before separation into subgroups shows unification of the symmetry groups as one expects in Grand Unified Theories (GUTS) thus fulfilling one type of unification.

Another, perhaps deeper, form of unification stems from the CASe groups of GiFT. We remember the CASe groups were the source of the fundamental symmetry group dimension array. (See chapter 6 in Appendix A.) The CASe groups were defined based on the need for General Relativistic transformations of sets of creation/annihilation operators between static and non-static coordinate systems. General Relativity is thus at the heart of Quantum Theory. The requirement of covariance in both dynamic equations and in creation/annihilation operator expressions completes the unification program for Quantum Theory and General Relativity.

CASe is required by General Relativistic transformations of Quantum creation/annihilation operators to non-static coordinate systems. The dimension arrays of the symmetry groups follow from CASe and give an initial unified symmetry group for each Hypercomplex Cosmology space. The Riemann-Christoffel curvature tensor embodies all the interactions including Gravitation and leads to a complete unified dynamics.. In the PseudoQuantum formulation its form is [4]

$$R'^{\beta}_{\sigma\nu\mu} = \sum_{\text{layers}} \{ R'_E{}^{1\beta}{}_{\sigma\nu\mu} + R'_E{}^{2\beta}{}_{\sigma\nu\mu} + R'_{SU(2)}{}^{1\beta}{}_{\sigma\nu\mu} + R'_{SU(2)}{}^{2\beta}{}_{\sigma\nu\mu} + R'_{DE}{}^{1\beta}{}_{\sigma\nu\mu} + R'_{DE}{}^{2\beta}{}_{\sigma\nu\mu} +$$

$$+ R'_{DSU(2)}{}^{1\beta}{}_{\sigma\nu\mu} + R'_{DSU(2)}{}^{2\beta}{}_{\sigma\nu\mu} + R'_{SU(3)}{}^{1\beta}{}_{\sigma\nu\mu} + R'_{SU(3)}{}^{2\beta}{}_{\sigma\nu\mu} + R'_U{}^{1\beta}{}_{\sigma\nu\mu} +$$

---

[3] Blaha (2021d) shows the Big Bang and subsequent expansion can be viewed as analogous to vacuum polarization. It shows free field behavior at the Big Bang.

[4] See chapters 22ff of Blaha (2018e).

$$+ R'_U{}^{2\beta}{}_{\sigma\nu\mu} + R'_V{}^{1\beta}{}_{\sigma\nu\mu} + R'_V{}^{2\beta}{}_{\sigma\nu\mu} \} + R'_B{}^{1\beta}{}_{\sigma\nu\mu} + R'_B{}^{2\beta}{}_{\sigma\nu\mu} + R^{1\beta}{}_{\sigma\nu\mu} + R^{2\beta}{}_{\sigma\nu\mu}$$

where

$$R'_E{}^{1\beta}{}_{\sigma\nu\mu} = ig^\beta{}_\sigma F_E{}^1{}_{\nu\mu}$$
$$R'_E{}^{2\beta}{}_{\sigma\nu\mu} = ig^\beta{}_\sigma F_{DE}{}^2{}_{\nu\mu}$$

$$R'_{DE}{}^{1\beta}{}_{\sigma\nu\mu} = ig^\beta{}_\sigma F_E{}^1{}_{\nu\mu}$$
$$R'_{DE}{}^{2\beta}{}_{\sigma\nu\mu} = ig^\beta{}_\sigma F_{DE}{}^2{}_{\nu\mu}$$

$$R'_{SU(2)}{}^{1\beta}{}_{\sigma\nu\mu} = ig^\beta{}_\sigma F_W{}^1{}_{\nu\mu}$$
$$R'_{SU(2)}{}^{2\beta}{}_{\sigma\nu\mu} = ig^\beta{}_\sigma F_{DW}{}^2{}_{\nu\mu}$$

$$R'_{DSU(2)}{}^{1\beta}{}_{\sigma\nu\mu} = ig^\beta{}_\sigma F_W{}^1{}_{\nu\mu}$$
$$R'_{DSU(2)}{}^{2\beta}{}_{\sigma\nu\mu} = ig^\beta{}_\sigma F_{DW}{}^2{}_{\nu\mu}$$

$$R'_{SU(3)}{}^{1\beta}{}_{\sigma\nu\mu} = ig^\beta{}_\sigma F_{SU(3)}{}^1{}_{\nu\mu}$$
$$R'_{SU(3)}{}^{2\beta}{}_{\sigma\nu\mu} = ig^\beta{}_\sigma F_{SU(3)}{}^2{}_{\nu\mu}$$

$$R'_U{}^{1\beta}{}_{\sigma\nu\mu} = ig^\beta{}_\sigma F_U{}^1{}_{\nu\mu}$$
$$R'_U{}^{2\beta}{}_{\sigma\nu\mu} = ig^\beta{}_\sigma F_U{}^2{}_{\nu\mu}$$

$$R'_V{}^{1\beta}{}_{\sigma\nu\mu} = ig^\beta{}_\sigma F_V{}^1{}_{\nu\mu}$$
$$R'_V{}^{2\beta}{}_{\sigma\nu\mu} = ig^\beta{}_\sigma F_V{}^2{}_{\nu\mu}$$

$$R'_B{}^{1\beta}{}_{\sigma\nu\mu} = ig^\beta{}_\sigma B^1{}_{\nu\mu}$$
$$R'_B{}^{2\beta}{}_{\sigma\nu\mu} = ig^\beta{}_\sigma B^2{}_{\nu\mu}$$

The unification of Quantum Theory and General Relativity is completed by the combination of space-time dimensions and internal symmetry representation dimensions in a dimension array in each HC space..

Lastly, the structure of the Hypercomplex Cosmology spaces spectrum follows directly from CASe. Thus the unification of Quantum Theory and General Relativity occurs on multiple levels.

FORM OF THE HYPERCOMPLEX SPACES SPECTRUM

Space Number o_s	Cayley-Dickson Number n	Cayley Number d_c	Dimension Array d_d	Space-time-Dimension r	CASe Group $su(2^{r/2}, 2^{r/2})$ CASe
0	10	1024	2048×2048	18	su(512,512)
1	9	512	1024×1024	16	su(256,256)
2	8	256	512×512	14	su(128,128)
3	7	128	256×256	12	su(64,64)
4	6	64	128×128	10	su(32,32)
5	5	32	64×64	8	su(16,16)
6	4	16	32×32	6	su(8,8)
7	**3**	**8**	**16×16**	**4**	**su(4,4)**
8	2	4	8×8	2	su(2,2)
9	1	2	4×4	0	su(1,1)

Figure 1.1. The NEW Hypercomplex Cosmology ten space spectrum. The space for our universe, is number 7, with Cayley-Dickson number 3 (which corresponds to octonions) is in bold type.

1.2 Structure of the Hypercomplex Cosmology Spectrum of Spaces

The "New View" Hypercomplex Spaces Spectrum in Fig. 1.1 reflects a number of points. The space-time dimensions r are set by starting at $r = 0$ and proceeding upward in steps of two dimensions to create a ten space spectrum. The choice of the ten space spectrum is essentially arbitrary but reasonable.

The increment steps are fixed at two because each odd space-time dimension has the same CASe group (and spinor array) as the next lower even number dimension, thus creating an unwanted duplication. Odd space-time dimensions are therefore excluded.

Given the choices of space-time dimensions the values in the other columns follows as we show in Appendix A and earlier books. In particular the spinor arrays, and more importantly, the CASe groups are shown to be composed of hypercomplex numbers ranging from Cayley-Dickson numbers $n = 1, \dots, 10$.

1.3 CASe Transformations

We have found that General Relativistic transformations from a static reference frame, such as the one generally thought to be that of our universe, to a non-static reference frame induces a transformation of creation/annihilation operators of the type of the $su(2^{r/2}, 2^{r/2})$ group for space-time dimension r. As shown in chapter 3 in Appendix A the creation/annihilation operator transformation[5] for a scalar field from coordinate system A with Fourier momentum α

---

[5] The equation numbering in this subsection is from chapter 3 in Appendix A.

$$\varphi_{1A}(x) = \Sigma_\alpha \, [a_{1\alpha} \, f_\alpha(x) + a^\dagger{}_{1\alpha} \, f_\alpha{}^*(x)] \qquad (3.11)$$
$$\varphi_{2A}(x) = \Sigma_\alpha \, [a_{2\alpha} f_\alpha(x) + a^\dagger{}_{2\alpha} \, f_\alpha{}^*(x)] \qquad (3.12)$$

to a coordinate system B with Fourier momentum β

$$\varphi_{1B}(x) = \Sigma_\beta \, [a_{1\beta} \, f_\beta(x) + a^\dagger{}_{1\beta} \, f_\beta{}^*(x)] \qquad (3.13)$$
$$\varphi_{2B}(x) = \Sigma_\beta \, [a_{2\beta} \, f_\beta(x) + a^\dagger{}_{2\beta} \, f_\beta{}^*(x)] \qquad (3.14)$$

where the B coordinate system creation/annihilation operators, can be expressed in terms of A coordinate system creation/annihilation operators, as

$$\varphi_{1B}(x) = \Sigma_\beta \, \Sigma_\alpha \, [(c_{11}a_{1\alpha} + c_{12}a_{2\alpha} + C_{11}a^\dagger{}_{1\alpha} + C_{12}a^\dagger{}_{2\alpha})f_\beta(x) +$$
$$+ \; (c'_{11}a^\dagger{}_{1\alpha} + c'_{12}a^\dagger{}_{2\alpha} + C'_{11}a_{1\alpha} + C'_{12}a_{2\alpha})f_\beta{}^*(x)] \qquad (3.19)$$
$$\varphi_{2B}(x) = \Sigma_\beta \, \Sigma_\alpha \, [(c_{21}a_{1\alpha} + c_{22}a_{2\alpha} + C_{21}a^\dagger{}_{1\alpha} + C_{22}a^\dagger{}_{2\alpha}) \, f_\beta(x) +$$
$$+ \; (c'_{21}a^\dagger{}_{1\alpha} + c'_{22}a^\dagger{}_{2\alpha} + C'_{21}a_{1\alpha} + C'_{22}a_{2\alpha})f_\beta{}^*(x)] \qquad (3.20)$$

where the c_{ij} and C_{ij}, and c'_{ij} and C'_{ij} are all functions of α.

Thus the A coordinate system creation/annihilation operators form a basis (an irreducible representation) of a CASe group (su(1, 1) in this example).

1.4 Hypercomplex Fundamental Representations of CASe Groups

The irreducible representations of the CASe groups are straight-forward. The su($2^{r/2}$, $2^{r/2}$) CASe group has an irreducible representation of $2^{r/2 + 2}$ real dimensions.

Turning to the fermion case for *our universe* (chapter 5 in Appendix A) we see the matrix representations of the su(4,4) group form a 16 × 16 fundamental representation in a 16 real dimension space. The space vectors have a form analogous to the set of creation/annihilation operators:

$$(b_{1\uparrow}, b_{2\uparrow}, b^\dagger{}_{1\uparrow}, b^\dagger{}_{2\uparrow}, \; b_{1\downarrow}, b_{2\downarrow}, b^\dagger{}_{1\downarrow}, b^\dagger{}_{2\downarrow}, \; d_{1\uparrow}, d_{2\uparrow}, d^\dagger{}_{1\uparrow}, d^\dagger{}_{2\uparrow}, \; d_{1\downarrow}, d_{2\downarrow}, d^\dagger{}_{1\downarrow}, d^\dagger{}_{2\downarrow}) \qquad (5.1)$$

where the arrows $\downarrow$ and $\uparrow$ specify spin. Note that the space has four quartets of vectors.[6] These operators correspond to a 16 real dimension su(4, 4) fundamental representation.

We now upgrade the discussion to *the irreducible representation of su(4, 4) over the sedenion numbers*. Physically we note that each of the 16 terms (taking account of spin) in the Fourier expansion of the fermion wave functions[7] begins with a single creation/annihilation operator:

$$\psi_{1A}(x) = \Sigma_{\alpha,s}[b_{1\alpha s}u_{\alpha s}f_\alpha(x) + d^\dagger{}_{1\alpha s}v_{\alpha s}f_\alpha{}^*(x)] \qquad (4.1)$$
$$\psi_{2A}(x) = \Sigma_{\alpha,s}[b_{2\alpha s}u_{\alpha s}f_\alpha(x) + d^\dagger{}_{2\alpha s}v_{\alpha s}f_\alpha{}^*(x)] \qquad (4.2)$$

plus Hermitean conjugates in a coordinate system A labeled with α.

---

[6] Note the form of eq. 5.1 appears to be complex-valued. But it is equivalent to a real-valued set.
[7] From chapter 4 in Appendix A.

Each creation/annihilation operator may be replaced with a sedenion since each term transforms to an expression such as the 16 term form of $b_{1\beta s'}$ taking account of the two values of s in the summation:

$$b_{1\beta s'} = \Sigma_{\alpha,x,s} (g_\beta, f_\alpha) u^\dagger_{\beta s'} u_{\alpha s} (c_{11s}b_{1\alpha s} + c_{12s}b_{2\alpha s} + C_{11s}b^\dagger_{1\alpha s} + C_{12}b^\dagger_{2\alpha s} + \qquad (4.26)$$
$$+ c'_{11s}d_{1\alpha s} + c'_{12s}d_{2\alpha s} + C'_{11s}d^\dagger_{1\alpha} + C'_{12s}d^\dagger_{2\alpha})$$

if we perform transformations with spin and b – d mixing.

Thus there are 16 sedenion vectors before and after a transformation and the group transformations matrix elements are promoted to 16×16 sedenions. We define an initial set of 16 sedenion vectors (each vector being a set of numeric labels representing creation/annihilation operators) for eqs. 4.1 – 4.2. They each have the form

$$S_i = (0, 0, \dots \overset{i^{th}}{y_i}, 0, 0, \dots) \qquad (5.4)$$

for i = 0, 1, 2, ..., 15 where y_i is the ith entry in eq. 5.1 above. Then we form a 16-vector

$$y = (S_0, S_1, \dots, S_{15}) \qquad (5.5)$$

with the wave functions now having the form

$$\psi_{1A}(x) = \Sigma_{\alpha,s}[S_{1\alpha s}u_{\alpha s}f_\alpha(x) + S^\dagger_{1\alpha s}v_{\alpha s}f_\alpha{}^*(x)] \qquad (1.1)$$
$$\psi_{2A}(x) = \Sigma_{\alpha,s}[S_{2\alpha s}u_{\alpha s}f_\alpha(x) + S^\dagger_{2\alpha s}v_{\alpha s}f_\alpha{}^*(x)] \qquad (1.2)$$

Defining y′ as the transformed set of values we have:

$$y' = Ty \qquad (1.3)$$

for transformation T, which embodies a General Relativistic transformation to a non-static reference frame. The transformation has a 16×16 matrix form with sedenion elements.

The generated y′ vector of 16 sedenions has each sedenion containing a value embodying the content of eq. 4.26 above transformed to a sedenion. We identify it as the form of the NEWQUeST dimension array.

The y′ vector becomes an array if its sedenions are expanded. The array specifies the set of dimensions of each irreducible representation of each symmetry group. It also specifies the set of fundamental fermions of NEWQUeST and NEWUST. See chapters 2a and 2b in Appendix A, which describes, and provides figures, for the features of NEWQUeST and NEWUST.

Thus we have a creation/annihilation transformation group su(4, 4) over the sedenion numbers specifying the symmetry dimension array and the fermion spectrum of NEWQUeST. This formulation will enable us to define transformations from a non-static Fundamental Reference Frame to a "normal" static reference frame where a base set of particles (or dimensions) is "expanded" by a transformation to the NEWQUeST/NEWUST set of particles (dimensions)

1.5 Fundamental Frame: Reduction of the sets of Symmetry Groups and Fermions to Smaller Numbers

In our universe we found 256 fundamental fermions in NEWQUeST/NEWUST and their precursors QUeST and UST. We now suggest that these fermions were generated from a non-static Fundamental Reference Frame with only 16 fermions (See Chapters 6 and 7 in Appendix A for more detail.)

We define 16 vectors of sedenions y in a non-static Fundamental Reference Frame. Each sedenion has a single fermion within it and has the form of eq. 5.4 above. We consider a General Relativistic transformation with corresponding transformation T in an *su(4, 4) irreducible representation over the sedenion numbers* that maps the non-static Fundamental Reference Frame to a static reference frame that has 256 fermions distributed in 16 output sedenions in a 16-vedtor y':

$$y' = Ty$$

which has the form

$$(\text{sedenion vector}) = T \, (\text{sedenion vector}) \qquad (1.3a)$$

The non-zero numeric values within y and y' are not significant–they only indicate the existence of a fermion. Thus the transformation T is physically not 1:1 with the General Relativistic transformation of coordinates. T represents one of a set of possible creation/annihilation transformations.

1.5.1 Fermions

There are 256 fermions generated from 16 fermions in the Fundamental Reference Frame. An examination of the form of the 256 fermion spectrum (Fig. 2b.5 in Appendix A) shows that it consists of 16 copies of a set of 16 fermions. Thus one can view T has generating a 16-furcation of the 16 Fundamental Frame fermions. At the Big Bang[8] we believe all particles were massless and so all 256 particles would be present until symmetry breaking was initiated. Thus the 16-furcation at the Big Bang point is not a problem.

1.5.2 Fermion Species

In chapters 6 and 7 in Appendix A we found the structure of the 16 fermions in the Fundamental Reference Frame. In chapter 2 below we find the CASe group for scalar particles. It is su(1, 1) in all spaces of Hypercomplex Cosmology (unlike the fermion case where the CASe group depends on space-time dimensions.) We now turn to consider the form of fermion species due to the appearance of two times in su(1, 1).[9] Just as the fermion CASe groups determine the dimension array, and consequently the

---

[8] See Blaha (2021d) which shows particles are quasi-free at the Big Bang.
[9] See chapter 5 for more information.

fundamental fermion spectrum, the su(1, 1) CASe group for scalar particles determines the form of the species of fermions in all Hypercomplex Cosmology spaces.

The su(1, 1) group has a metric with two real time coordinates:

$$ds^2 = t_{01}{}^2 + t_{02}{}^2 - x_1{}^2 - x_2{}^2 \tag{1.4}$$

In Blaha (2007b) we showed that the four species of fermions: e-type, ν-type, q-up-type and q-down-type followed from the four types of boosts of the Lorentz Group (O(1, 3), which is often expressed as SL(2, **C**)). Complex Lorentz group boosts separate into sublight, superluminal, complex sublight, and complex superluminal boosts yielding the four species respectively.

Now we have a more interesting situation with two time coordinates.[10] As subsection 5.4.1 of chapter 5 describes: one light speed divides the set of possible fermions into two "superspecies", namely normal and Dark fermions. The second light speed divides each superspecies into four parts. The result is the fermion species *in our universe* of Fig. 1.2, which includes both Normal and Dark sectors. It is also the form of fermion species in the other Hypercomplex Cosmology spaces.

The 16 fermions of the Fundamental Reference Frame can be separated into the 16 species of Fig. 1.2 with account taken of the occurrence of quark species as triplets.

The 16-furcation of the 16 Fundamental Frame fermions then yields the NEWQUeST/NEWUST fermion spectrum of Fig. 1.2 in Appendix A.

A complete, detailed spectrum of fundamental fermions in our universe, and other universes, emerges, part of which appears in the Standard Model for our universe.

t$_{01}$:
| | sublight | | | | superluminal | | |
| | **Normal** | | | | **Dark** | | |

t$_{02}$:

sublight		superluminal		sublight		superluminal	
e-type	q-up-type	ν-type	q-down-type	e-type	q-up-type	ν-type	q-down-type

Figure 1.2. Separation of fermion species due to su(1, 1) and its two times. The number of corresponding fermions is 16 if account is taken of quarks appearing as triplets. The occurrence of quark in triplets is motivated by the need for 16 fermions. Thus the separation by two light speeds may account for the Standard Model form of the fermion species and also for triplets of quarks. Experimentally, superluminal neutrinos appear to move at the speed of light due to their negligible masses. Down-type quarks, being confined within particles, do not directly display their superluminal nature. Dark matter, being unobserved is superluminal. Their superluminal nature may be part of the cause for their

---

[10] Subsection 5.4.1 describes the transformation between a Fundamental Frame where superluminal quantities are present and a static frame where they are transformed to sublight quantities.

current unobservability. It is also possible that the superluminal nature of these quantities may "wash out" when a transformation is made from non-static to static reference frames.

1.5.3 Dimensions – Symmetries

The symmetries in the Fundamental Reference Frame also have irreducible representations totaling to 16 real dimensions. Again the su(1, 1) CASe group with its two time dimensions structures the set of groups. Fig. 1.3 displays the separation of the representations due to sublight vs. superluminal parts.

t_{01}:	sublight		superluminal	
	U(4)		U(4)	
t_{02}:	sublight	superluminal	sublight	superluminal
	O (1, 3)	SU(2)⊗U(1)	U(1)	SU(3)
Irreducible Representation Real Dimensions	4	4	2	6

Figure 1.3. Separation of symmetries due to su(1, 1) and its two times. The number of dimensions is 16. The separation by two light speeds may account for the Standard Model form of the interaction groups. Superluminal SU(3) may account, in part, for color confinement, which remains somewhat of a mystery. The superluminal symmetries can appear like sublight symmetries in terms of their experimental impact. The SU(2)⊗U(1) symmetry superluminality is submerged by the extremely small neutrino masses. It is also possible that the superluminal nature of these quantities may "wash out" when a transformation is made from non-static to static reference frames.

The symmetries in Fig. 1.3 and the fermions in Fig. 1.2 for the Fundamental Reference Frame form the ingredients of a one dimension Standard Model. The transformation of the ingredients yields the NEWQUeST fermion spectrum and the NEWQUeST symmetries spectrum occupying 256 dimensions. The 16 dimension symmetries of Fig. 1.3 are 16-furcated to generate the Normal and Dark symmetry spectrum of NEWQUeST and NEWUST displayed in Fig. 2b.4. Eight of the duplicates are each U(4)⊗U(4), and the other eight are each O(1, 3)⊗SU(2)⊗U(1)⊗U(1)⊗SU(3). At the Big Bang universe creation point the vector bosons of all symmetries are massless and quasi-free as pointed out earlier.

We transform the eight copies of O(1, 3) contained within the 16 copies into seven U(2) Connection groups that define interactions among the Normal and Dark sectors, and the four layers within them. Fig. 2b.3 of Appendix A shows the roles of these groups.

1.6 Fundamental Reference Frames for Other Hypercomplex Cosmology Spaces

The previous sections have described the generation of fermions and symmetries for our universe. In this section we describe the generation of fermions and symmetries in other Hypercomplex Cosmology spaces from CASe groups. We also describe the Fundamental Reference Frames for each space. The CASe groups of Hypercomplex Cosmology spaces are listed in Fig. 1.1 shown below.

We note the n level CASe group is $su(2^{n-1}, 2^{n-1})$, the total real dimensions of the CASe group irreducible representation is $2^{n+1} \times 2^{n+1} = 2^{2n+2}$, the space-time dimension is $r = 2n - 2$, and the total dimensions in the symmetry array of the Hypercomplex Cosmology space is 2^{2n+2}. See Fig. 1.1.

FORM OF THE HYPERCOMPLEX SPACES SPECTRUM

Space Number O_S	Cayley-Dickson Number n	Cayley Number d_c	Dimension Array d_d	Space-time-Dimension r	CASe Group $su(2^{r/2},2^{r/2})$ CASe
0	10	1024	2048×2048	18	su(512,512)
1	9	512	1024×1024	16	su(256,256)
2	8	256	512×512	14	su(128,128)
3	7	128	256×256	12	su(64,64)
4	6	64	128×128	10	su(32,32)
5	5	32	64×64	8	su(16,16)
6	4	16	32×32	6	su(8,8)
7	**3**	**8**	**16 × 16**	**4**	**su(4,4)**
8	2	4	8×8	2	su(2,2)
9	1	2	4×4	0	su(1,1)

Figure 1.1. The NEW Hypercomplex Cosmology ten space spectrum. The space for our universe, is number 7, with Cayley-Dickson number 3 (which corresponds to octonions) is in bold type.

Following the development of chapters 6 and 7 in Appendix A we first note equal size of the CASe transformation matrix and the fermion and symmetry dimension arrays as shown in Fig. 7.1.

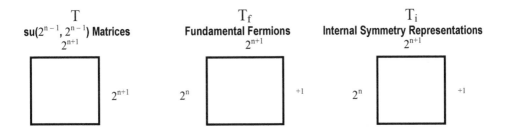

Figure 7.1   Comparison of arrays in CASe and for the Hypercomplex Cosmology space with Cayley-Dickson number n.

We assume a non-static level n Fundamental Reference Frame exists, from which a General Relativistic transformation generates the Hypercomplex Cosmology symmetry group and fermion spectrum of a static reference frame with the transformation(s):

$$F_{fermions} = T'F_{fundamental} \tag{7.1}$$
$$I_{QUeST} = T''I_{fundamental} \tag{7.2}$$

having the form

$$(n+1)\text{-ion vector} = T\ (n+1)\text{-ion vector} \tag{1.5}$$

We assume $T' = T''$. Then $F_{fundamental}$ would be a 2^{n+1} component set of vectors (each with 2^{n+1} elements) in an irreducible $su(2^{n-1}, 2^{n-1})$ representation over the level n Cayley-Dickson numbers of the form

$$\overset{i^{th}}{S_i = (0, 0, \ldots y_i, 0, 0, \ldots)} \tag{1.6}$$

for $i = 0, 1, 2, \ldots, 2^{n+1} - 1$ where y_i is the ith entry in the hypercomplex number.
We form a real-valued 2^{n+1}-vector

$$F_{fundamental} = (S_0, S_1, \ldots, S_k) \tag{1.7}$$

(where $k = 2^{n+1} - 1$) to represent the Fundamental Reference Frame hypercomplex vectors. Each fermion hypercomplex number in $F_{fundamental}$ has one primitive fermion within it. Each symmetry dimension hypercomplex number in $I_{fundamental}$ has one dimension within it

After applying the transformations of eqs. 7.1 and 7.2 above, a 2^{n+1} component vector of "filled" 2^{n+1}-vectors appears: for $F_{fermions}$ and I_{QUeST}. The 2^{n+1} fermions of the Fundamental Reference Frame become 2^{n+1} filled vectors of fermions giving a total of 2^{2n+2} fermions in the static reference frame.

Also the 2^{n+1} dimensions of the Fundamental Reference Frame are 2^{n+1}filled vectors of dimensions giving a total of 2^{2n+2} dimensions in the static reference frame. It can be viewed as a $2^{n+1} \times 2^{n+1}$ real-valued dimension array for the nth space of Hypercomplex Cosmology.[11]

The reference frame containing the reduced sets of fermions and dimensions is called the hypercomplex number *Fundamental Frame*. The General Relativistic fermion transformation is symbolized in Fig. 7.2 of Appendix A and shown below.

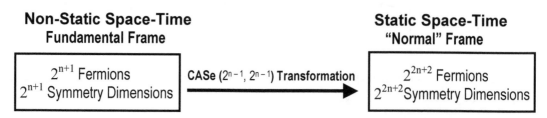

Figure 7.2. The CASe transformation from a non-static Fundamental Frame to a "Normal" static space-time reference frame with fermions and dimensions both increasing by a factor of 2^{n+1}.

1.6.1 Fermions within the Fundamental Frame

The 2^{2n+2} fermions in the output static frame consist of 2^{n+1} copies of the 2^{n+1} fermions in the Fundamental Reference Frame. The 2^{n+1} fermions in the Fundamental Reference Frame consist of 2^{n+1} vectors of the form of eq. 1.6. These vectors transform to 2^{n+1} filled 2^{n+1}-vectors. The total fermion count is 2^{2n+2} fermions.

The 2^{n+1} fermions in the Fundamental Reference Frame consist of 2^{n-3} copies of the 16 fermions in Fig. 1.2 above. We define the number of copies in the Fundamental Frame as

$$n_c = 2^{n+1}/16 = 2^{n-3} \tag{1.8}$$

Each copy of the fermions has different fermions in it. For example for n = 3, three of the copies of the 16 copies, are for the known three generations of fundamental fermions.

1.6.2 Symmetry Groups within the Fundamental Frame

There are 2^{n+1} symmetry dimensions within the Fundamental Frame 2^{n+1}-vector.[12] Each of the 2^{n+1} vectors in the Fundamental Reference Frame contains one of the 2^{n+1} symmetry group dimensions within a 2^{n+1}-vector of the form of eq. 5.4 above.

The 2^{2n+2} dimensions in the output static frame consist of 2^{n+1} copies of the 2^{n+1} dimensions in the Fundamental Reference Frame. The 2^{n+1} dimensions in the Fundamental Reference Frame are viewed as transformed to $n_c = 2^{n-3}$ copies of the 16 dimensions represented by symmetry groups in Fig. 1.3. Each copy of the set of n_c copies are for a different set of internal symmetries.

---

[11] These results are equivalent to those of Octonion Cosmology described in previous work.

[12] in the irreducible $su(2^{n-1}, 2^{n-1})$ group representation over the n-ion numbers.

1.7 Meaning of Space-time in Hypercomplex Cosmology

Each space-time in Hypercomplex Cosmology plays an important role. The question arises: How does the number of dimensions in a space-time arise? In this section we answer that question and show a direct connection to the Fundamental Frames. We begin by noting

$$r = 2n - 2$$
$$d_d = 2^{n+1} \times 2^{n+1} = 2^{2n+2}$$
$$d_r = 2^{n+1}$$

(1.9)

where r is the dimension of the space-time, d_d is the number of elements in the dimension array for space n and d_r is the number of rows (and columns) in the dimension array.

These equations imply

$$n = \log_2 (d_r/2)$$

(1.10)

and thus

$$r = \log_2 (d_d/16)$$

(1.11)

or, in words, r is the logarithm (base 2) of the total number of copies of the set of fermions in Fig. 1.2 within the fermion array = 2^{2n-2} copies of the Fig. 1.2 set of fermions. Note the d_d dimensions in the dimension array include the dimensions of all the internal symmetry irreducible representations and the space-time dimensions.

Similarly r is also the logarithm (base 2) of the total number of copies of the set of symmetries in Fig. 1.3 determined by the dimension array = 2^{2n-2} copies of the Fig. 1.3 set of symmetries.

Thus the space-time dimension of each space is determined ultimately by its Fundamental Frame fermions and internal symmetries. The number of space-time dimensions in a space, the total number of internal symmetry dimensions of a space, and the number of fundamental fermions of the space's Fundamental Frame are linked thus supporting the reality of the Fundamental Frame.

This chapter has shown how the sets of fermions and dimensions are generated by a CASe transformation due to a General Relativistic transformation of the fermions and dimensions from the Fundamental Reference Frame to a static reference frame.

2. New Formulation of *Scalar* Space Particles

In previous work we defined space particles based on spinor arrays. The spinor arrays were taken to be the source of the dimension arrays of the spaces of Octonion Cosmology. While this approach may still be viable we have been led to consider a new approach based on spaces generated by CASe transformations of creation/annihilation operators appearing in Fourier expansions of PseudoQuantum fields. The form of the CASe group transformations is quite similar, and better for our purposes than the form of spinor arrays previously studied.

2.1 Scalar Space Particle Case

We begin with eqs. 3.11 – 3.20 appearing in chapter 1. A scalar field in reference frame A has the form

$$\varphi_{1A}(x) = \Sigma_\alpha \, [a_{1\alpha} \, f_\alpha(x) + a^\dagger_{1\alpha} f_\alpha^*(x)] \qquad (3.11)$$
$$\varphi_{2A}(x) = \Sigma_\alpha \, [a_{2\alpha} f_\alpha(x) + a^\dagger_{2\alpha} f_\alpha^*(x)] \qquad (3.12)$$

where α represents the Fourier momentum. In coordinate system B with Fourier momentum β

$$\varphi_{1B}(x) = \Sigma_\beta \, [a_{1\beta} \, f_\beta(x) + a^\dagger_{1\beta} f_\beta^*(x)] \qquad (3.13)$$
$$\varphi_{2B}(x) = \Sigma_\beta \, [a_{2\beta} f_\beta(x) + a^\dagger_{2\beta} f_\beta^*(x)] \qquad (3.14)$$

where the B coordinate system creation/annihilation operators can be expressed in terms of A coordinate system creation/annihilation operators, as[13]

$$\varphi_{1B}(x) = \Sigma_\beta \, \Sigma_\alpha \, [(c_{11}a_{1\alpha} + c_{12}a_{2\alpha} + C_{11}a^\dagger_{1\alpha} + C_{12}a^\dagger_{2\alpha})f_\beta(x) +$$
$$+ \, (c'_{11}a^\dagger_{1\alpha} + c'_{12}a^\dagger_{2\alpha} + C'_{11}a_{1\alpha} + C'_{12}a_{2\alpha})f_\beta^*(x)] \qquad (3.19)$$

$$\varphi_{2B}(x) = \Sigma_\beta \, \Sigma_\alpha \, [(c_{21}a_{1\alpha} + c_{22}a_{2\alpha} + C_{21}a^\dagger_{1\alpha} + C_{22}a^\dagger_{2\alpha}) \, f_\beta(x) +$$
$$+ \, (c'_{21}a^\dagger_{1\alpha} + c'_{22}a^\dagger_{2\alpha} + C'_{21}a_{1\alpha} + C'_{22}a_{2\alpha})f_\beta^*(x)] \qquad (3.20)$$

where the c_{ij} and C_{ij}, and c'_{ij} and C'_{ij} are all functions of α and possibly x.

We now introduce a new notation that is well adapted to the CASe group formulation. The fields in reference frame A can be put in a "matrix" form using the vector:

---

[13] Note the transformation to the B coordinate system mixes the type 1 and type 2 fields.

$$Q_A(x, \alpha) = \begin{bmatrix} f_\alpha(x)\, a_{1\alpha} \\ f_\alpha{}^*(x)\, a^\dagger{}_{1\alpha} \\ f_\alpha(x)\, a_{2\alpha} \\ f_\alpha{}^*(x)\, a^\dagger{}_{2\alpha} \end{bmatrix} \tag{2.1}$$

and similarly for the fields in reference frame B.

The fields in A can be expressed by

$$\varphi_{1A}(x) = \Sigma_\alpha\, [1, 1, 0, 0] Q_A(x, \alpha)$$
$$\varphi_{2A}(x) = \Sigma_\alpha\, [0, 0, 1, 1] Q_A(x, \alpha)$$

Next we remove the dependence on x of the A fields using the A frame momentum operator $P_A{}^\mu$ applied to each of the four components of Q_A individually.

$$Q_A(0, \alpha) = \exp(iP_A{}^\mu x_\mu) Q_A(x, \alpha) \exp(-iP_A{}^\mu x_\mu) \tag{2.1}$$

where 0 denotes a value of x reducing $f_\alpha(x)$ to a real-valued constant, which we choose to be 1. Then

$$Q_A(0, \alpha) = \begin{bmatrix} a_{1\alpha} \\ a^\dagger{}_{1\alpha} \\ a_{2\alpha} \\ a^\dagger{}_{2\alpha} \end{bmatrix} \tag{2.2}$$

We are now able to use the machinery of the su(1,1) CASe group to implement General Relativistic transformations to non-static reference frames with CASe operator transformations of the form of eqs. 3.19 and 3.20 above. Applying a CASe transformation T to Q_A we obtain

$$Q_B(0, \alpha) = T(\alpha) Q_A(0, \alpha)$$

$$= \begin{bmatrix} c_{11}a_{1\alpha} + c_{12}a_{2\alpha} + C_{11}a^\dagger{}_{1\alpha} + C_{12}a^\dagger{}_{2\alpha} \\ c'_{11}a^\dagger{}_{1\alpha} + c'_{12}a^\dagger{}_{2\alpha} + C'_{11}a_{1\alpha} + C'_{12}a_{2\alpha} \\ c_{21}a_{1\alpha} + c_{22}a_{2\alpha} + C_{21}a^\dagger{}_{1\alpha} + C_{22}a^\dagger{}_{2\alpha} \\ c'_{21}a^\dagger{}_{1\alpha} + c'_{22}a^\dagger{}_{2\alpha} + C'_{21}a_{1\alpha} + C'_{22}a_{2\alpha} \end{bmatrix} \tag{2.3}$$

We can then obtain eqs. 3.19 and 3.20 by inserting $f_\beta(x)$ and $f_\beta{}^*(x)$ factors in each component of $Q_B(0, \alpha)$. However it is preferable to use $\exp(-iP_A{}^\mu x_\mu)$ to restore coordinate dependence:

$$Q_B(x, \alpha) = \exp(-iP_A{}^\mu x_\mu) Q_B(0, \alpha) \exp(iP_A{}^\mu x_\mu)$$
$$= \exp(-iP_A{}^\mu x_\mu) T(\alpha) \exp(iP_A{}^\mu x_\mu) \exp(-iP_A{}^\mu x_\mu) Q_A(0, \alpha) \exp(-iP_A{}^\mu x_\mu)$$

$$= \exp(-iP_A{}^\mu x_\mu)T(\alpha) \exp(iP_A{}^\mu x_\mu) Q_A(x, \alpha)$$
$$= T(x, \alpha) Q_A(x, \alpha) \qquad (2.4)$$

The CASe transformation $T(\alpha)$ acquires a coordinate dependence making it local in coordinate space x and in "momentum" space α.

Note 1: The momentum operator $P_A{}^\mu$ must be expressed in terms of creation/annihilation operators for use in eq. 2.4 in a fashion analogous to CASe transformations as shown in section 3.1 in Appendix A.

Note 2: The momentum operator $P_A{}^\mu$ and Q_A creation/annihilation operator expressions can induce parallel c-number formulations.

The $T(\alpha)$ transformation is also expressed in terms of creation and annihilation operators as we did in section 3.1 in Appendix A and earlier in Blaha (2021i) and (2021j). Then the operator $T(x, \alpha)$ has $f_\beta(x)$ and $f_\beta{}^*(x)$ factors in the creation and annihilation operators within it.

$Q_A(0, \alpha)$ furnishes the basis of a su(1, 1) space. It also induces a c-number numeric array representation that gives the basis for a 4×4 dimension array and for a similar 4×4 array of fermions. The dimension array corresponds to a zero space-time dimension Hypercomplex Cosmology space for Cayley-Dickson number n = 1.

The 4×4 dimension space induced by su(1, 1) is implicit in the scalar space particles. The space has no size (a point) and no mass-energy initially. However we shall see later that the annihilation of a fermion-antifermion pair can inelastically produce a pair of scalar space particles that each contain mass-energy in a Big Bang state initially. Each space can then undergo expansion similar to that of our universe.[14]

Note *the su(1, 1) CASe space is independent of the space-time dimension* of the fields. There are four components of Q_A in eq. 2.2 in any space-time dimension.

2.2 Scalar Space Particle Field with a SU(N) Symmetry

We now consider a scalar space particle field with an SU(N) symmetry. We define the particles as in the previous section, but with an SU(N) index:

Frame A:
$$\varphi^{Nk}_{1A}(x) = \Sigma_\alpha [a^{Nk}_{1\alpha} f_\alpha(x) + a^{Nk}_1{}^\dagger{}_\alpha f_\alpha{}^*(x)] \qquad (2.5)$$
$$\varphi^{Nk}_{2A}(x) = \Sigma_\alpha [a^{Nk}_{2\alpha} f_\alpha(x) + a^{Nk}_2{}^\dagger{}_\alpha f_\alpha{}^*(x)] \qquad (2.6)$$

Frame B:
$$\varphi^{Nk}_{1B}(x) = \Sigma_\beta [a^{Nk}_{1\beta} f_\beta(x) + a^{Nk}_1{}^\dagger{}_\beta f_\beta{}^*(x)] \qquad (2.7)$$
$$\varphi^{Nk}_{2B}(x) = \Sigma_\beta [a^{Nk}_{2\beta} f_\beta(x) + a^{Nk}_2{}^\dagger{}_\beta f_\beta{}^*(x)] \qquad (2.8)$$

where k = 1, 2, ... , N.

Then following the same procedure as in the preceding section we find

---

[14] See Blaha (2021d) for a derivation of universe expansion using a particle framework based on a vacuum polarization analogue.

$$Q^N_A(x, \alpha) = \begin{bmatrix} f_\alpha(x) \, a^{N1}_{1\alpha} \\ f_\alpha{}^*(x) \, a^{N1}_1{}^\dagger{}_\alpha \\ f_\alpha(x) \, a^{N1}_{2\alpha} \\ f_\alpha{}^*(x) \, a^{N1}_2{}^\dagger{}_\alpha \\ \cdots \\ f_\alpha(x) \, a^{NN}_{1\alpha} \\ f_\alpha{}^*(x) \, a^{NN}_1{}^\dagger{}_\alpha \\ f_\alpha(x) \, a^{NN}_{2\alpha} \\ f_\alpha{}^*(x) \, a^{NN}_2{}^\dagger{}_\alpha \end{bmatrix} \qquad (2.9)$$

The creation/annihilation operator representation has been "multiplied" N-fold. Consequently the CASe group is su(N, N). The corresponding real-valued dimension array has the form $d_d = 4N \times 4N$. See Fig. 1.1.

The $4N \times 4N$ dimension space induced by su(N, N) is implicit in the SU(N) scalar space particles. The space has no size and no mass-energy initially. However later we shall see that the annihilation of a fermion-antifermion pair can inelastically produce a pair of SU(N) scalar space particles that each contain mass-energy in a Big Bang state initially. Each space then undergoes expansion similar to that of our universe.[15]

Note the su(N, N) CASe space is independent of the space-time dimension of the fields. There are $4N$ components of Q^N_A for any space-time dimensions.

The procedure followed (as in eq. 2.4) maintains the vacuum properties of the original $\varphi^k_{1A}(x)$ and $\varphi^k_{2A}(x)$ fields and their creation/annihilation operators:

$$\begin{aligned} Q^N_B(x, \alpha) &= \exp(-iP_A{}^\mu x_\mu) Q^N_B(0, \alpha) \exp(iP_A{}^\mu x_\mu) \\ &= \exp(-iP_A{}^\mu x_\mu) T^N(\alpha) \exp(iP_A{}^\mu x_\mu) \exp(-iP_A{}^\mu x_\mu) Q^N_A(0, \alpha) \exp(-iP_A{}^\mu x_\mu) \\ &= \exp(-iP_A{}^\mu x_\mu) T^N(\alpha) \exp(iP_A{}^\mu x_\mu) \, Q^N_A(x, \alpha) \\ &= T^N(x, \alpha) \, Q^N_A(x, \alpha) \end{aligned} \qquad (2.10)$$

2.3 Hypercomplex Cosmology Form of Scalar Fields

Scalar fields can be defined for each Hypercomplex Cosmology space with Cayley-Dickson number n. Starting from

$$\varphi^{Nk}_{1A}(x) = \Sigma_\alpha \, [a^{Nk}_{1\alpha} \, f_\alpha(x) + a^{Nk}_1{}^\dagger{}_\alpha \, f_\alpha{}^*(x)] \qquad (2.11)$$

$$\varphi^{Nk}_{2A}(x) = \Sigma_\alpha \, [a^{Nk}_{2\alpha} f_\alpha(x) + a^{Nk}_2{}^\dagger{}_\alpha \, f_\alpha{}^*(x)] \qquad (2.12)$$

we define

$$\varphi^{Nk}_{1Ar'rM}(x) = \Sigma_\alpha \, [a^{Nk}_{1\alpha} \, f_\alpha(x) + a^{Nk}_1{}^\dagger{}_\alpha \, f_\alpha{}^*(x)] O(N, r, M) \qquad (2.13)$$

$$\varphi^{Nk}_{2A\,r'rM}(x) = \Sigma_\alpha \, [a^{Nk}_{2\alpha} f_\alpha(x) + a^{Nk}_2{}^\dagger{}_\alpha \, f_\alpha{}^*(x)] \, O(N, r, M) \qquad (2.14)$$

---

[15] See Blaha (2021d) for a derivation of universe expansion using a particle framework based on a vacuum polarization analogue.

where O(N, r, M) symbolizes the *implicit* CASe group space of unitary group SU(N), r is the space-time dimension of the implicit space, and M is the mass-energy of the implicit space. Initially M = 0. The space-time of the space containing the scalar fields $\varphi^{Nrk}_{iA\,r'}(x)$ for i = 1, 2 is r', which is not equal to r in general.

For the nth level in the Hypercomplex Cosmology spectrum we set

$$N = 2^{n-1} \qquad (2.15)$$

$$r = 2n - 2 \qquad (2.16)$$

to specify a space of level n, where r is its space-time dimension. Fermions of the $(n+1)^{th}$ level will annihilate to produce scalar particles containing an nth level space O(N, r, M). See chapter 4.

The space-time (dimension r') of the x coordinates will turn out to be different from the space-time dimension r within the space O when we consider space generation by fermion-antifermion annihilation in chapter 4. The scalar fields of Cayley-Dickson number n will have a dimension field of size $d_d = 2^{n+1} \times 2^{n+1}$. (See Fig. 1.1 reproduced below) They have N and r set by eqs. 2.15 and 2.16. The CASe group is su(N, N) = su(2^{n-1}, 2^{n-1}).

FORM OF THE HYPERCOMPLEX SPACES SPECTRUM

Space Number O_s	Cayley-Dickson Number n	Cayley Number d_c	Dimension Array 2^{n+1} d_d	Space-time-Dimension r	CASe Group $su(2^{r/2},2^{r/2})$ CASe
0	10	1024	2048×2048	18	su(512,512)
1	9	512	1024×1024	16	su(256,256)
2	8	256	512×512	14	su(128,128)
3	7	128	256×256	12	su(64,64)
4	6	64	128×128	10	su(32,32)
5	5	32	64×64	8	su(16,16)
6	4	16	32×32	6	su(8,8)
7	**3**	**8**	**16 × 16**	**4**	**su(4,4)**
8	2	4	8×8	2	su(2,2)
9	1	2	4×4	0	su(1,1)

Figure 1.1. The NEW Hypercomplex Cosmology ten space spectrum. The space for our universe, is number 7, with Cayley-Dickson number 3 (which corresponds to octonions) is in bold type. This spectrum, which is based on CASe spaces analysis, eliminates non-regularities in the spectrum, based on spinor spaces, in Appendix A.

3. New Formulation of *Fermion* Space Particle Theory

We now turn to develop the Quantum Field Theory of fermion Space Particles. In section 1.6 we determined the CASe groups of the Hypercomplex Cosmology spaces. The fermion CASe groups are determined in part by the relation between space-time dimensions and the spin states of fermions. The number of spin states of a fermion n_s in an r dimension space-time is:

$$n_s = 2^{r/2-1} \tag{3.1}$$

For our 4D universe there are two spin states. Each of the spin states has a corresponding set of creation/annihilation operators. (See sections 1.3, 1.4 and 1.6.)

From eq. 5.1 in section 1.4 we see that there are eight operators for each spin state. For spin up we have

$$(b_{1\uparrow}, b_{2\uparrow}, b^{\dagger}{}_{1\uparrow}, b^{\dagger}{}_{2\uparrow},\ d_{1\uparrow}, d_{2\uparrow}, d^{\dagger}{}_{1\uparrow}, d^{\dagger}{}_{2\uparrow}) \tag{3.2}$$

Thus for n_s states we find the number of creation/annihilation operators n_o is

$$n_o = 8n_s = 8*2^{r/2-1} \tag{3.3}$$

and thus the CASe group is

$$su(n_c, n_c) \tag{3.4}$$

where

$$n_c = n_o/4 \tag{3.5}$$

as one can see from Fig. 1.1 reproduced below.

In section 1.6 we noted the level[16] n CASe group is $su(2^{n-1}, 2^{n-1})$, the total dimensions of the group irreducible representation is $2^{n+1} \times 2^{n+1} = 2^{2n+2}$, the space-time dimension is $r = 2n - 2$, and the dimension of the symmetry array of the Hypercomplex Cosmology space is 2^{2n+2}. Consequently

$$n_o = 2^{n+1} \tag{3.6}$$

and

$$n_c = 2^{n-1} \tag{3.7}$$

as confirmed by Fig. 1.1 below.

---

[16] Cayley-Dickson number n.

NEW FORM OF THE HYPERCOMPLEX SPACES SPECTRUM

Space Number O_s	Cayley-Dickson Number n	Cayley Number d_c	Dimension Array d_d	Space-time-Dimension r	CASe Group $su(2^{r/2}, 2^{r/2})$ CASe
0	10	1024	2048 × 2048	18	su(512,512)
1	9	512	1024 × 1024	16	su(256,256)
2	8	256	512 × 512	14	su(128,128)
3	7	128	256 × 256	12	su(64,64)
4	6	64	128 × 128	10	su(32,32)
5	5	32	64 × 64	8	su(16,16)
6	4	16	32 × 32	6	su(8,8)
7	**3**	**8**	**16 × 16**	**4**	**su(4,4)**
8	2	4	8 × 8	2	su(2,2)
9	1	2	4 × 4	0	su(1,1)

Figure 1.1. The NEW Hypercomplex Cosmology ten space spectrum. The space for our universe, is number 7, with Cayley-Dickson number 3 (which corresponds to octonions) is in bold type.

The fermion fields in level n of the Hypercomplex Cosmology spectrum with space-time dimension $r = 2n - 2$ are

$$\psi^n_{1ArM}(x) = \Sigma_{\alpha,s}[b_{1\alpha s}u_{\alpha s}f_\alpha(x) + d^\dagger_{1\alpha s}v_{\alpha s}f_\alpha^*(x)] \, O(2^{n+1}, r, M) \qquad (3.8)$$
$$\psi^n_{2A\,rM}(x) = \Sigma_{\alpha,s}[b_{2\alpha s}u_{\alpha s}f_\alpha(x) + d^\dagger_{2\alpha s}v_{\alpha s}f_\alpha^*(x)] \, O(2^{n+1}, r, M) \qquad (3.9)$$

where $O(2^{n+1}, r, M)$ symbolizes the *implicit* CASe group space $su(2^{n-1}, 2^{n-1})$ with a symmetry dimension array containing 2^{2n+2} elements, r is the space-time dimension of the implicit space, and M is the mass-energy of the implicit space. Initially M = 0. Note the space-time of the n space has the same number of dimensions as the implicit space O space-time dimensions.

4. Generation of Space Instances by Fermion-Antifermion Annihilation

We consider the generation of the sequence of spaces of Hypercomplex Cosmology from an ultimate space (God-space) of 18 space-time dimensions. The generation proceeds through a fermion-antifermion annihilation in a "parent" space of level n that creates a pair of scalar space particles with each containing a "child" sister space of level n − 1 that is populated with a mass-energy M. We assume the creation process is inelastic. It conserves spatial momentum in the parent space. Energy conservation takes place with the transfer of mass-energy to the child spaces.

The interaction term is:

$$g\psi^n_{2A\,r'M_1}(x)\;\psi^n_{2A\,r'M_2}(x)\;\varphi^{Nk}_{2A\,r'rM_3}(x)\;\varphi^N_{k2A\,r'rM_4}(x) \tag{4.1}$$

with a summation over k, and where

$$N = 2^{n-2} \tag{4.2}$$
$$r = 2n - 4 \tag{4.3}$$

specifies that the scalar space O is one level below the space of level n. Note the PseudoQuantum type 2 fields used for interactions are

$$\psi^n_{2A\,r'M}(x) = \Sigma_{\alpha,s}[b_{2\alpha s}u_{\alpha s}f_\alpha(x) + d^\dagger_{2\alpha s}v_{\alpha s}f_\alpha*(x)]\;O(2^{n+1}, r, M) \tag{4.4}$$
$$\varphi^{Nk}_{2A\,rM}(x) = \Sigma_\alpha\,[a^{Nk}_{2\alpha}f_\alpha(x) + a^{Nk}_2{}^\dagger_\alpha f_\alpha*(x)]\;O(N, r, M) \tag{4.5}$$

The interaction applies to fermions which are in a level n space with $r' = 2n - 2$ space-time dimensions and an overall set of 2^{2n+2} dimensions for symmetries and space-time.

The scalar bosons also exist in the level n space. They contain an implicit space symbolized by O(N, r, M) where N and r are given by eqs. 4.2 and 4.3 for a space with Cayley-Dickson number n − 1. Symbolically we represent the transition of eq. 4.1 with

$$O(2^{n+1}, 2n - 2, M_1)\;O(2^{n+1}, 2n\text{ - }2, M_2) \rightarrow O(2^{n-2}, 2n \text{ - } 4, M_3)\;O(2^{n-2}, 2n - 4, M_4) \tag{4.6}$$

by eq. 4.2. The two produced sister spaces are of level n − 1. Thus we say that fermion-antifermion annihilation in level n generates two sister spaces (universes) of the next level down of level n − 1.

Note: the space-time dimensions in the eq. 4.1 interaction are related by

$$r = r' \text{ - } 2 \tag{4.7}$$

consistent with the creation of a space in the n − 1 level of the Hypercomplex Cosmology spectrum of spaces.

We may take $M_1 = M_2 = 0$ initially. The interaction may be inelastic with a transfer of mass-energy to the scalar particle spaces with M_3 and M_4 non-zero in general. We view the interaction as analogous to colliding "mud balls" where energy of motion is transferred to the resultant mud balls.

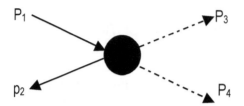

Figure 4.1. The diagram corresponding to the interaction in eq. 4.1.

The spatial momentum is conserved:

$$\mathbf{P_1} + \mathbf{P_2} = \mathbf{P_3} + \mathbf{P_4} \tag{4.8}$$

The energy is conserved if account is taken of the transfer of energy to the space within each mud ball. With the energy transfer the generated spaces can then evolve by expansion as we see in our own universe.[17]

$$E_1 + M_1 + E_2 + M_2 = E_3 + M_3 + E_4 + M_4 \tag{4.9}$$

4.1 Hypercomplex Cosmology Spectrum

Starting from the n = 10 space with 18 space-time dimensions we can generate the remainder of the 10 spaces through fermion-antifermion annihilation processes as depicted in Fig. 4.2. We assume the n = 10 space is a universe containing mass-energy that supports the creation process. (It would be possible to have the n = 10 space generated by a precursor space.)

The sequence of spaces are nested. There are several chains of spaces due to the stage by stage generation of sister spaces. Fig. 2a.4 in Appendix A shows a hierarchy of spaces generated through fermion-antifermion annihilation.

4.2 Sister Universes

The creation of sister universes might account for some unusual asymmetries in our universe such as the observed excess of Dark matter over Normal matter.

---

[17] See Blaha (2021d).

4.3 Space O features

The implicit spaces symbolized by O(N, r, M) may be susceptible to further study. A created space O begins with a non-zero mass-energy M. We take it to be a point in the parent space. It has the momentum and energy of the scalar space particle containing it in the parent space.[18] It has symmetries and a space-time of dimension r. It proceeds to evolve from the creation point by an expansion fueled by mass-energy M. See Blaha (2021d) for a model of space expansion. We defer the further study of O to a future publication.

---

[18] Blaha (2018e) discusses the dynamics of universes moving through a Megaverse (Multiverse).

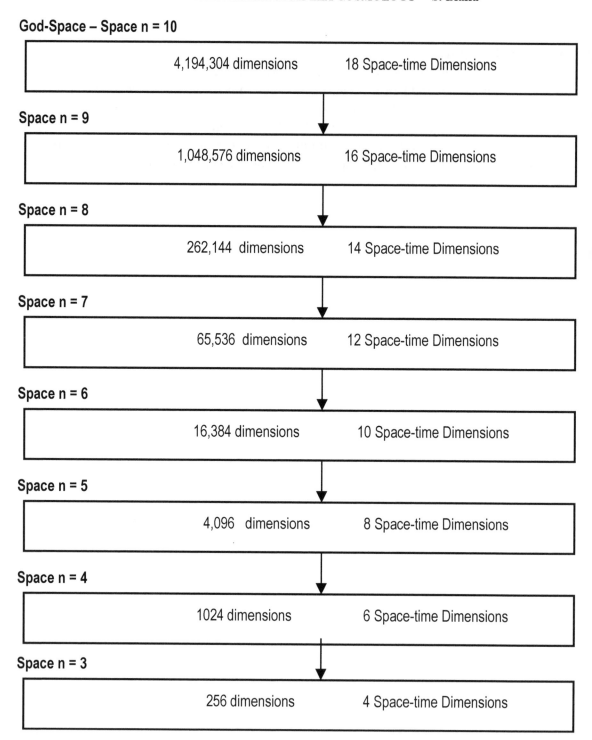

God-Space – Space n = 10

| 4,194,304 dimensions | 18 Space-time Dimensions |

Space n = 9

| 1,048,576 dimensions | 16 Space-time Dimensions |

Space n = 8

| 262,144 dimensions | 14 Space-time Dimensions |

Space n = 7

| 65,536 dimensions | 12 Space-time Dimensions |

Space n = 6

| 16,384 dimensions | 10 Space-time Dimensions |

Space n = 5

| 4,096   dimensions | 8 Space-time Dimensions |

Space n = 4

| 1024 dimensions | 6 Space-time Dimensions |

Space n = 3

| 256 dimensions | 4 Space-time Dimensions |

Figure 4.2. A sequence of fermion-antifermion annihilations generating instances in the set of Hypercomplex Cosmology spaces leading to our universe.

5. Origin of Symmetries and the Fermion Spectrum

This chapter describes the origin of the forms of symmetries and of the form of the fundamental fermion spectrum at the creation point of a space (universe) before the dynamical evolution of the space occurs.

Over the past fifty years or so many attempts have been made to understand the fairly odd looking symmetries and fundamental fermions of the Standard Model. It appears that Hypercomplex Cosmology, when developed from the GiFT CASe groups formalism of creation/annihilation operators, provides a complete explanation of the symmetries of the Standard Model (and NEWUST and NEWQUeST) and the generations form of fundamental fermions including the 3-fold nature of quarks.

The approach pioneered here avoids a commitment to an arbitrary symmetry group or a GUT. The symmetry groups appear naturally. Ultimately they emanate from the simple creation/annihilation structure of a quantum field, which is based on four creation/annihilation operator sets.

5.1 Initial State of a Space

When a space is created within a scalar space particle, its form is based on multiple copies of creation/annihilation operators:

$$Q_A(0, \alpha) = \begin{bmatrix} a_{1\alpha} \\ a^{\dagger}_{1\alpha} \\ a_{2\alpha} \\ a^{\dagger}_{2\alpha} \end{bmatrix} \qquad (2.2)$$

In the space of Cayley-Dickson number n there are $N = 2^{n-1}$ copies giving a real dimension of $4*2^{n-1} = 2^{n+1}$ for the irreducible representation of $su(2^{n-1}, 2^{n-1})$.

Thus the $su(2^{n-1}, 2^{n-1})$ fundamental representation matrices have an inherent separation into 4×4 subblocks. There are 2^{2n-2} subblocks in the level n dimension array of size $2^{n+1} \times 2^{n+1}$. For example the level 3 dimension array contains 16 subblocks within its 256 dimensions.

The 16 element subblocks are of critical importance for determining the structure of the symmetry groups and also for determining the form of the fermion spectrum.

5.2 Origin of Fermion Spectrum

Each 16 element subblock of the set of 2^{2n-2} subblocks represents 16 fermions, which, looking ahead to section 5.5, take the form

	Normal					Dark		
e-type	q-up-type	v-type	q-down-type		e-type	q-up-type	v-type	q-down-type

$\qquad (5.1)$

There are 2^{2n-2} rows of fermions of the above form. If we consider the example of n = 3, which corresponds to our universe, we see there are 16 rows. We subdivide the rows into four layers of four generations and discover the form of the fermion spectrum in the Unified SuperStandard Theory (UST) (and also QUeST), The UST is described in Blaha (2018e). It contains Fig. 2b.5 reproduced below showing the four layers of four generations.

The fermion spectrum for other values of n can also be put in a multi-layered, multi-generation form.

The overall structure of the fermion spectrum is thus obtained for all spaces of Hypercomplex Cosmology. The one row form of eq. 5.1 is derived in section 5.4.

5.3 Origin of Symmetries' Structure

The dimension array for symmetries also has a 16 element subblock structure. Each 16 element subblock within the total set of 2^{2n-2} subblocks represents 16 symmetry dimensions allocated to various symmetries. In section 5.5 we show that each symmetry subblock can be put in the form of 32 dimension sets of two rows. Some sets are for Normal matter interactions and some are for Dark matter interactions.

$$\textbf{U(4)} \qquad\qquad\qquad \textbf{U(4)} \qquad\qquad (5.2)$$

$$\textbf{O (1, 3)} \qquad \textbf{SU(2)} \otimes \textbf{U(1)} \qquad \textbf{U(1)} \qquad \textbf{SU(3)} \qquad (5.3)$$

The subblock structure applies to all spaces of Hypercomplex Cosmology. Thus the pattern of symmetries in eqs. 5.2 and 5.3 is present throughout Hypercomplex Cosmology.

Again turning to our universe where n = 3 we find there are $2^{2n-3} = 8$ subblocks. We allocate four subblocks to Normal matter interactions and four subblocks to Dark matter interactions. Thus each of the four layers of Normal matter has a set of symmetries specified by eqs. 5.2 and 5.3. Similarly each of the four layers of Dark matter has a set. Again this result agrees with NEWUST and NEWQUeST. See Figs. 5.1 and 5.2 below. See also Fig. 2b.4 below from Blaha (2018e).

The Fermion Periodic Table (n = 3)

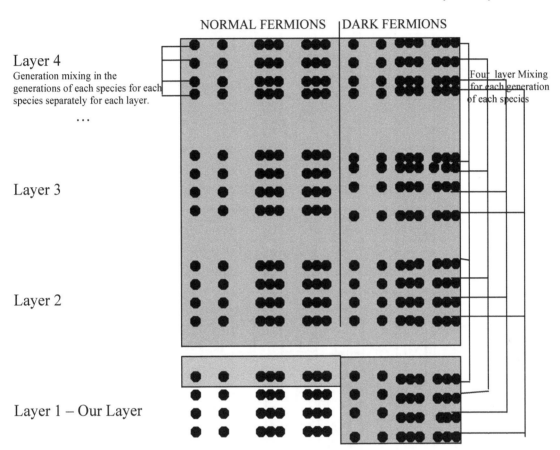

Figure 2b.5. Fermion particle spectrum and partial examples of the pattern of mass mixing of the Generation group and of the Layer group. Unshaded parts are the known fermions with an additional, as yet not found, 4th generation. The lines on the left side (only shown for one layer) display the Generation mixing within each layer. The Generation mixing occurs within each layer using a separate Generation group for each layer. The lines on the right side show Layer group mixing (for Dark matter) with the mixing among all four layers for each of the four generations individually. There are four Layer groups for Normal matter and four Layer groups for Dark matter.. There are 256 fundamental fermions. QUeST and UST have the same fermion spectrum. Each row is a 16 fermion subblock.

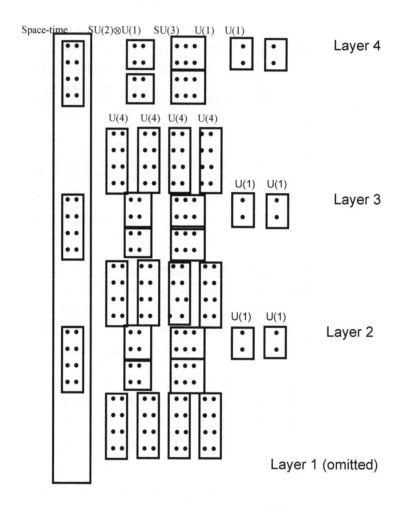

Figure 2b.4. Three of the four layers of QUeST internal symmetry groups (and space-time) for Cayley-Dickson space 4. Layer 1 which has an identical form was omitted due to "page space" limitations. Note the left column of blocks combine to specify a 4 dimension octonion space-time. Note each layer has 64 dimensions.

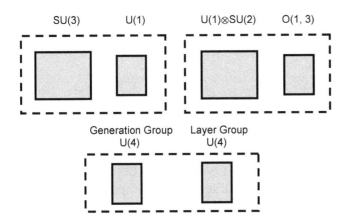

Figure 5.1 (Modified slightly from Fig. 3.1a of Blaha (2021c)) An SU(3)⊗U(1) The set of rows in eqs. 5.2 and 5.3 consisting of dimension 16 subblocks with each encircled by a "dotted" line.

	Normal				Dark		
O (1, 3)	SU(2)⊗U(1)	U(1)	SU(3)	O (1, 3)	SU(2)⊗U(1)	U(1)	SU(3)
	U(4) U(4)				U(4) U(4)		
O (1, 3)	SU(2)⊗U(1)	U(1)	SU(3)	O (1, 3)	SU(2)⊗U(1)	U(1)	SU(3)
	U(4) U(4)				U(4) U(4)		
O (1, 3)	SU(2)⊗U(1)	U(1)	SU(3)	O (1, 3)	SU(2)⊗U(1)	U(1)	SU(3)
	U(4) U(4)				U(4) U(4)		
O (1, 3)	SU(2)⊗U(1)	U(1)	SU(3)	O (1, 3)	SU(2)⊗U(1)	U(1)	SU(3)
	U(4) U(4)				U(4) U(4)		

Figure 5.2 The four layers of symmetry groups for our universe (n = 3). The symmetry groups are all independent of each other. Their dimensions total to 256. This set of symmetry groups is the same as UST and QUeST except that 7 O(1, 3) groups are transformed to seven U(2) Connection groups. See Fig. 2b.3 in Appendix A.

5.4 Origin of Fermion Species (Types)

In chapters 2 and 3 we found the structure of the 16 fermions in the Fundamental Reference Frame. In chapter 2 we found the CASe group for scalar particles. It is su(1, 1) in all spaces of Hypercomplex Cosmology (unlike the fermion case where the CASe group depends on the number of spins for a space-time dimension.)

We now turn to consider the form of fermion species due to the appearance of two time coordinates in su(1, 1). Just as the fermion CASe groups determine the dimension array, and consequently the fundamental fermion spectrum, the su(1, 1) CASe group for scalar particles determines the form of the species of fermions in all Hypercomplex Cosmology spaces.

The su(1, 1) group has a metric with two real time coordinates:

$$ds^2 = t_{01}{}^2 + t_{02}{}^2 - x_1{}^2 - x_2{}^2 \tag{1.4}$$

In Blaha (2007b) we showed that the four species of fermions: e-type, ν-type, q-up-type and q-down-type followed from the four types of boosts of the Lorentz Group (often expressed as SL(2, **C**)). The speed of light boost separated complex Lorentz group boosts into sublight, superluminal, complex sublight, and complex superluminal yielding the four species respectively.

Now we have a more interesting situation with two time coordinates. As chapter 1 describes: one light speed divides the set of possible fermions into two "superspecies", namely normal and Dark fermions. The second light speed divides each superspecies into four parts as in the preceding paragraph. The result is the fermion species *in our universe* of Fig. 1.2, which includes both Normal and Dark sectors.

The 16 fermions of the Fundamental Reference Frame can be separated into the 16 species of Fig. 1.2 with account taken of the occurrence of quark species as triplets.

The 16 "copies" of the 16 Fundamental Frame fermions then yields the NEWQUeST/NEWUST fermion spectrum of Fig. 1.2 in Appendix A.

A complete, detailed spectrum of fundamental fermions in our universe emerges, part of which appears in the Standard Model for our universe and the other spaces of Hypercomplex Cosmology.

5.4.1 Fundamental Frame Space-time

The space-time of the Fundamental frame must be non-static. It is due to a General Relativistic transformation from a static space-time. If the transformation is complex-valued then a metric of the form of eq. 1.4 above can be a *local*[19] metric of the Fundamental Frame. Since there are two time coordinates this metric is non-static.

In this frame one may define sublight and superluminal motion. Upon transformation back to a static reference frame:

Non-static Frame $\rightarrow$ static Frame

the fermions and symmetries are transformed back to the "sublight" fermions and symmetries of Hypercomplex Cosmology.

---

[19] Generally a "Lorentz-like" frame can be locally defined in any space-time.

5.5 Origin of Symmetries' Species

The symmetries in the Fundamental Reference Frame also have irreducible representations totaling to 16 real dimensions. Again the su(1, 1) CASe group with its two time dimensions structure the set of groups. Fig. 1.3 below displays the separation of the representations due to sublight vs. superluminal parts.

t_{01}:

	sublight		superluminal	
	Normal		**Dark**	

t_{02}:

sublight	superluminal	sublight	superluminal
e-type | q-up-type	v-type | q-down-type	e-type| q-up-type	v-type| q-down-type

Figure 1.2. Separation of fermion species due to su(1, 1) and its two times. The number of corresponding fermions is 16 if account is taken of quarks appearing as triplets. The occurrence of quark in triplets is motivated by the need for 16 fermions. Thus the separation by two light speeds may account for the Standard Model form of the fermion species and also for triplets of quarks. Experimentally, the superluminal neutrinos appear to move at the speed of light due to their negligible masses. Down-type quarks, being confined within particles, do not directly display their superluminal nature. Dark matter, being unobserved, may be superluminal. Their superluminal nature may be part of the cause for their current unobservability. Or, the superluminal aspect in the Fundamental Frame may "wash out" when transformed to the static frame that we observe.

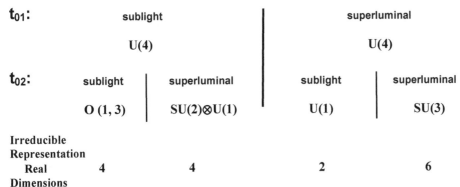

t_{01}:

	sublight		superluminal
	U(4)		U(4)

t_{02}:

sublight	superluminal	sublight	superluminal
O (1, 3)	SU(2)⊗U(1)	U(1)	SU(3)

Irreducible Representation Real Dimensions			
4	4	2	6

Figure 1.3. Separation of symmetries due to su(1, 1) and its two times. The number of dimensions is 16. The separation by two light speeds may account for the Standard Model form of the interaction groups. Superluminal SU(3) may account, in part, for color confinement, which remains somewhat of a mystery. The superluminal symmetries can appear like sublight symmetries in terms of their experimental impact. The SU(2)⊗U(1) symmetry may be submerged by the extremely small neutrino masses. Or ,the superluminal aspect in the

Fundamental Frame may "wash out" when transformed to the static frame that we observe.

5.6 One Generation Standard Model in the Fundamental Frame

The symmetries in Fig. 1.3 and the fermions in Fig. 1.2 for the Fundamental Reference Frame form the ingredients of a one generation Standard Model. A General Relativistic transformation of the ingredients yields the NEWQUeST fermion spectrum and the NEWQUeST symmetries spectrum occupying 256 dimensions. The 16 dimensions symmetries of Fig. 1.3 are duplicated 16 times to generate the Normal and Dark symmetry spectrum of NEWQUeST and NEWUST displayed in Fig. 2b.4. Eight of the duplicates are $U(4){\otimes}U(4)$, and the other eight (which can be viewed as originating from symmetry breakdown) are each $O(1, 3){\otimes}SU(2){\otimes}U(1){\otimes}U(1){\otimes}SU(3)$. At the Big Bang universe creation point the vector bosons of all symmetries are massless and quasi-free as pointed out earlier.

We transform the eight copies of $O(1, 3)$ contained within the 16 copies into seven $U(2)$ Connection groups that define interactions among the Normal and Dark sectors, and the four layers within them. Fig. 2b.3 of Appendix A shows the roles of these groups.

5.7 The Fundamental Reference Frame

The Fundamental Reference Frame is a non-static reference frame. It appears that many such frames may have the reduced spectrum of fermions and the reduced set of symmetries, and may have the "two time" metric of eq. 1.4 described in subsection 5.4.1. The dynamics of a non-static frame are convoluted due to the absence of a well-defined time variable or the presence of multiple time coordinates.

We regard the Fundamental Frame to be generated at the starting point of a space created in a fermion-antifermion annihilation.

Higgs particles for symmetry breaking may also be present in the Fundamental Frame and the corresponding static frame.

6. The Fundamental Appearance of Four

An examination of the structure of Hypercomplex Cosmology shows that the number 4 and its multiples appear throughout the theories. If we seek the source of its appearance we find it follows from the number of boson and fermion creation/annihilation operators appearing in quantum field operators using the GiFT and CASe formalisms. The PseudoQuantum operators in a scalar field number four or multiples of four if a symmetry is present. The PseudoQuantum operators in a fermion field number eight (a multiple of four), or multiples of eight due to the number of spin states in higher space-time dimensions.

In a conventional, non-PseudoQuantum formulation multiples of two would follow. Multiples of two are present in multiples of four. Thus multiples of two are also present in the PseudoQuantum formulation. However *odd* multiples of two are not present as can be seen in chapters 2 through 5.

Thus the appearance of multiples of four in the derivations of fermion species and symmetry species argues for the validity of GiFT's CASe formulation.

Appendix A. Unified SuperStandard Theory, Octonion Cosmology, GiFt and CASe

This appendix is an extract from Blaha (2021i).

1. Unification of General Relativity and Quantum Theory

The unification of General Relativity and Quantum Theory has been a goal of Physics for approximately one hundred years. The meaning, and process, of unification has been the subject of discussion for almost as long. This book formulates a unification procedure and then proceeds to examine the ingredients and form of unification. We will see that Octonion Cosmology contains the essential ingredients for unity: a combination of space-times with internal symmetries in each of its universes.

The form of General Relativity, in relation to the form of other particle interactions, reveals a unity based on GiFT. A careful analysis of elementary particle field theory in differing coordinate systems leads to the Generalized Field Theory's (GiFT) Creation/Annihilation Space (CASe), which furnishes a further basis for unification with General Relativity.

1.1 The Form of General Relativity

General Relativity was first formulated by Einstein based on the Classical Physics (differential equations) of coordinate systems. It appeared different from the gauge field formulation of internal symmetries. As a result there seemed to be a gap between General Relativity and Elementary Particle Theory. Attempts were made to bridge the gap by quantizing General Relativity directly. These attempts seem to have not changed the view that General Relativity and Elementary particle theory were inherently different.

A different approach within a *vierbein* framework appeared due to the work of T. Kibble and others. This approach led to a gauge field formulation of General Relativity Thus this type of formulation of General Relativity places it on the same basis as the formulation of internal symmetry gauge fields.

A significant issue for unification is the appearance of infinities in gravitation perturbation theory. These infinities do not appear when Gravitation perturbation theory is done within the framework of the Two-Tier Theory within GiFT.

1.2 Octonion Cosmology Framework

The similarity between the *vierbein* formulation of General Relativity and elementary particle internal symmetry gauge theory raises the question of origin: Is there a common origin?

Octonion Cosmology provides a common origin in each of its ten universes (spaces). Part of each space has coordinates for each fundamental irreducible representation. Part of each space is for space-time coordinates. The transformation properties of the space-times lead to a General Relativity for each universe. Both the General Relativity and the internal symmetries can be formulated as gauge theories. Are there relations between coupling constants? Yes, but they differ in principle from universe to universe. Is there an ultimate unification value for all coupling constants in a universe? There may be such a unification value. But one must also consider the partition of internal symmetries in each universe due to the original form of spinor spaces, and in hypercomplex number multiplication tables, as we described in previous work. Partitioning may obviate the value of a common coupling constant value.

1.3 Quantization Process

The process of quantization begins with the issue of particle quantization vs. wave quantization: Should matter and energy be quantized as particles or as waves? The answer is particle quantization is simple if one wishes creation and annihilation events. Particles can interact to produce other particles in a "billiard ball" picture. Waves require complex differential equations at best.

Given a particle formulation it makes sense to define particle creation, a, and annihilation operators, $a^\dagger$, that can embody interactions that can change a set of incoming particles to a set of outgoing particles. At this point, if we introduce a "momentum" α for each particle, the question arises: Can particles have the same momentum? The Pauli Exclusion Principle states that fermion particles cannot. So we can set $(a^\dagger)^2 = 0$ for fermions. Bosons are not constrained by the Pauli Exclusion Principle so $(a^\dagger)^2 \neq 0$. One concludes the particle definition concepts by defining field operators for particles with Fourier expansions in α.

Having defined fields for particles in a coordinate system, then we note that a particle in that coordinate system becomes a superposition of particles when viewed in another coordinate in general. Einstein encountered this problem of exploring dynamic equations in one coordinate system in comparison to their equivalent in other coordinate systems. He resolved the problem with General Relativity.

1.4 Quantization in Differing Coordinate Systems

In our earlier work we showed that transformations between coordinate systems, where one or both coordinate systems do not have a Killing vector, are questionable in ordinary quantum field theory. We then showed quantization via GiFT provides a clear definition of particle states and their relation in the respective coordinate systems. (We pursue the GiFT approach further for bosons and fermions later in this book.)

The GiFT formulation of field theory is based on transformations between coordinate systems as is General Relativity. Thus both General Relativity and GiFT support the unity of General Relativity and Quantum Theory. General Relativity performs for coordinate systems what GiFT does for quantization in coordinate systems. Fig. 1.1 shows the correspondence between General Relativistic coordinate transformations and GiFT transformations in the space of creation/annihilation operators. A General Relativistic transformation between coordinate systems with Killing vectors does not induce a GiFT transformation. A General Relativistic transformation between coordinate systems where one coordinate system has a Killing vector, and the other does not, does induce a non-trivial GiFT transformation.

As we pointed out in earlier work GiFT embodies Quantum Theory in its entirety: PseudoQuantum Theory, Two-Tier Theory, Quantum Field Theory, and Quantum Mechanics. GiFT supports CQ Mechanics which contains Quantum Mechanics and Classical Mechanics. Thus we see the unity in General Relativity and Quantum Theory.

1.5 GiFT Creation/Annihilation Spaces

GiFT has creation/annihilation spaces created with CASe transformations. In section 3.2 and in chapter 4 we show that creation/annihilation operators can form a basis for a 16-dimension space, a CASe space, which has a metric that is invariant under su(4,4) transformations for fermion particles, and su(n,m) in general for fermion particles in higher dimensions where n and m depend on the space-time dimension.

The structure of CASe spaces and their group transformations parallels that of Special Relativity and General Relativity in space-time.

Thus General Relativity and GiFT Quantum Theory, which contains Quantum Field Theory and Quantum Mechanics, closely parallel each other in the features of transformations between coordinate systems, and in the gauge field format of interactions.

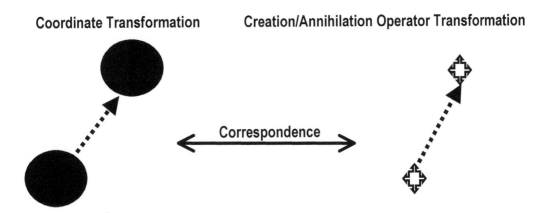

Coordinate Transformation   **Creation/Annihilation Operator Transformation**

Correspondence

Figure 1.1.   General Coordinate Transformations parallel Creation/Annihilation Operator Transformations.

1.6 Unification

The above considerations show the unity of General Relativity and Quantum Theory in our formulation. We find a unifying basis for General Relativity and Quantum Theory in considerations of transformations between coordinate systems, and in the existence of the relation between space-time transformations and CASe transformations. Vierbein gravity and the local gauge theory of internal symmetries exhibit a similar formulation. The structure of an overall fully unified theory of General Relativity and GiFT is evident.

2a. Pioneer's Progress in Octonion Cosmology

Recent years have witnessed a new effort to develop a comprehensive, deeper theory of Elementary Particles, Cosmology, and Quantum Theory. In this chapter we epitomize the developments of the Unified SuperStandard Theory and the Octonion Cosmology that embodies it. Octonion Cosmology requires a deeper quantum formulation that we have provided in GiFT,[20] which contains our PeseudoQuantum Theory and our Two-Tier Theory. Together these efforts open a much wider framework beyond the four space-time dimensions of the Standard Model and the Cosmological Standard Model. The developments are based on features of Spinor spaces that provide a new arena for the exploration of dynamical and group features of Elementary Particles and the Cosmos.

This chapter, and the following chapter 2b, outline the theory of Octonion Cosmology and its connection to the Unified SuperStandard Theory (NEWUST) through NEWQUeST. It is based on the author's books: *Quantum Space Theory With Application to Octonion Cosmology & Possibly To Fermionic Condensed Matter* and *Beyond Octonion Cosmology*, and *Beyond Octonion Cosmology II* as well as other previous books by the author. The discussion will be in a narrative form to make it accessible to a wider range of readers. Readers familiar with these books may proceed directly to chapter 3.

2a.1 Fundamental Required Features

Certain features are necessary for a satisfactory modern theory of fundamental Physics. These features include:

1. Space-times within which dynamic processes can take place.
2. A satisfactory Quantum formalism – GiFT.
3. Matter.
4. Energy.
5. A space of internal symmetries for SU(3) and so on.

In addition to these necessary requirements are a viable Quantum Field Theory, and a satisfactory Quantum Space Theory[21] of fermion and boson quantum fields containing subspaces, and describing the generation and features of these spaces.

2a.2 The Primary Space of Octonion Cosmology

Octonion Cosmology begins with the definition of a primary space,[22] which we will call space 0, from which the other spaces evolve. Space 0 has a twenty dimension

---

[20] Blaha (2021i).
[21] See Blaha (2021g).
[22] The space corresponds to Cayley number 2048. See Fig. 1.1 or Blaha (2021c) for details

space-time (nineteen spatial dimensions and one time dimension.) It contains $2^{22} - 20 = 4,194,284$ additional dimensions for internal symmetries such as SU(2)⊗U(1) and SU(3) symmetries.

A space is a universe. Each universe, after space 0, has a space-time with dimensions ranging from zero to eighteen. In addition it has dimensions for irreducible representations of the internal symmetry groups that it supports. Examples of groups are SU(3) – the Strong interactions and SU(2)⊗U(1) – the ElectroWeak interactions. We will use the terms space and universe interchangeably.

We view space 0 as initially having a certain size in both space and time. In that space-time a fermion has a *spinor array* with row and column widths of 1024×1024 and thus 1,048,576 entries. We assume a fermion(s) exists in the 20 space-time dimension space 0.

We now contract the space-time to a 20 dimension point. The fermion spinor array size (rows and columns) remains the same and generates a spinor array – a *spinor space* with $1024^2 = 1,048,576$ dimensions.[23] This space has Cayley-Dickson number n = 10 in Fig. 2a.1. This space is a universe.

THE TEN OCTONION SPACES SPECTRUM

Octonion Space Number O_s	Cayley-Dickson Number n	Cayley Number d_c	Dimension Array Total d_d	Space-time-Dimension r	Fermion Spinor Array Total d_s
0	10	1024	1024 × 1024	18	512 × 512
1	9	512	512 × 512	16	256 × 256
2	8	256	256 × 256	14	128 × 128
3	7	128	128 × 128	12	64 × 64
4	6	64	64 × 64	10	32 × 32
5	5	32	32 × 32	8	16 × 16
6	**4**	**16**	**16 × 16**	$6 \rightarrow 4^{24}$	**8 × 8**
7	3	8	8 × 8	4	4 × 4
8	2	4	4 × 4	2	2 × 2
9	1	2	2 × 2	0	1 × 1

Figure 2a.1. The Octonion Cosmology ten space spectrum. The space for our universe, number 6, (Cayley-Dickson number 4) is in bold type.

2a.3 Space 10

A fermion in the 20 space-time dimension space 0 contains a spinor space (space 10) populated with matter and energy due to its mass. The n = 10 spinor space, with its elementary particles and energies, is the space[25] of a sub-universe. It has an 18

---

[23] The transformation of a fermion spinor array to a dimension array is described in detail in the case of fermion annihilation in chapter 6 of Blaha (2021g).

[24] The space-time dimensions become 4 as in the Unified SuperStandard Theory (UST) through either transfer of dimensions to internal symmetries or by the compactification of two dimensions. As a result the pattern of fermion-antifermion annihilations producing spaces n = 7, 8, and 9 is disrupted.

[25] See Fig. 1.1.

dimension space-time. The choice of 18 dimensions is motivated by the goal of having a Cayley-Dickson spectrum of spaces. Later we will find a deeper reason for this choice. See chapters 5 and 6.

In space n = 10 a fermion-antifermion annihilation can take place. It can be visualized with the following Feynman-like diagram:[26]

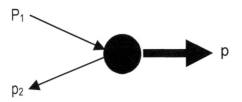

where p_1 and p_2 are the momentum of the fermion and antifermion respectively and p is the momentum of the produced space-containing particle. The produced (scalar) space particle, which exists in space 10,[27] contains a spinor array/space that is the space of another universe of the type of space n = 9 in Fig. 2a.1.

Space n = 9, with its mass and energy, is created using the mass and energy from the annihilation in space 10. It has $512^2 = 262,144$ dimensions, of which 16 are space-time dimensions.

This space-time is the first of the series of nine spaces (universes) generated through fermion-antifermion annihilations. They produce a Cayley-Dickson number based spectrum of spaces. See Figs. 2a.1 and 2a.2 below.

2a.4 The Further Generation of Spaces

The spectrum of spaces (universes) is generated through a series of fermion-antifermion annihilations in space after space. The Cayley-Dickson pattern of spaces emerges from the choice of the number of space-time dimensions in each space. We descend from space to space by decreasing the space-time dimension by two at each space.[28] The below Fig. 2a.3 shows the general pattern of nesting of space instances in a nested form.

The result of the sequence of "annihilations" is the spectrum of Fig. 2a.1, which corresponds to the Cayley-Dickson numbered spaces from 10 through 1 with space-time dimensions from 18 through 0. See Fig. 2a.5.

In this iterative process there can be any number of subspace instances generated from a space instance by multiple fermion-antifermion annihilations. For example see Fig. 2a.4.

---

[26] We assume off shell fermions.

[27] Note the produced particle has a mass and momentum in space 10.

[28] This choice is motivated by deeper reasons tied to General Relativity. See chapters 5 and 6.

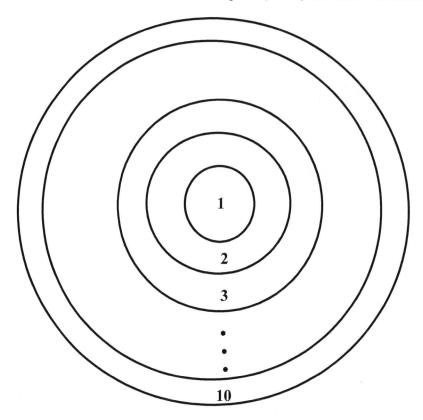

Figure 2a.2. Numbered form of the ten space instances of Octonion Cosmology. They are generated by fermion-antifermion annhilations. The outermost space instance (Cayley-Dickson space 10) is generated first from the 20 space-time dimension space. The Space 9 instance is generated within space 10 and so on.

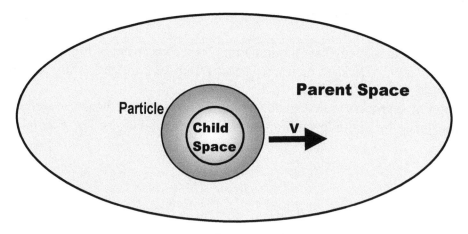

Figure 2a.3. The general pattern of a (parent) space containing a particle of velocity **v** that has an internal space instance (child) of lower dimension and

lower Cayley-Dickson number. The space is itself within an instance of a higher space. The spectrum's spaces form a nested sequence. Each space contains the next lower space within it. The instances of spaces also form a nested sequence as shown in Fig. 1.2 below.

A SMALL COSMOS

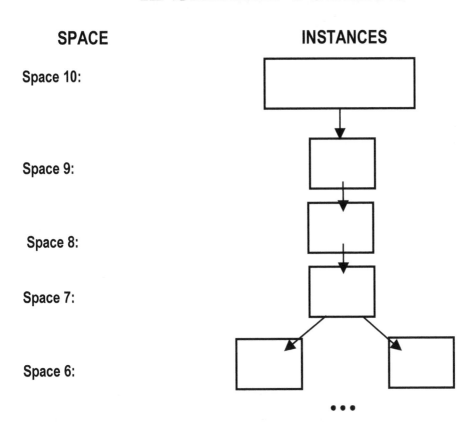

Figure 2a.4. A hierarchy of space instances one "sibling" and three "cousin" universes. *The entire hierarchy resides in space number 10.*

God-Space – Space 0: Complex Octonion Octonion Octonion

1,048,576 dimensions $2^{20/2} = 1024$ rows/columns	18 Space-time Dimensions	512 512-spinors

Space 1: Octonion Octonion Octonion

262,144 dimensions $2^{18/2} = 512$ rows/columns	16 Space-time Dimensions	256 256-spinors

Space 2: Quarternion Octonion Octonion

65,536 dimensions $2^{16/2} = 256$ rows/columns	14 Space-time Dimensions	128 128-spinors

Space 3: Complex Octonion Octonion

16,384 dimensions $2^{14/2} = 128$ rows/columns	12 Space-time Dimensions	64 64-spinors

Space 4: Octonion Octonion

4,096 dimensions $2^{12/2} = 64$ rows/columns	10 Space-time Dimensions	32 32-spinors

Space 5: Quaternion Octonion

1,024 dimensions $2^{10/2} = 32$ rows/columns	8 Space-time Dimensions	16 16-spinors

Space 6: Complex Octonion

256 dimensions $2^{8/2} = 16$ rows/columns	4 Space-time Dimensions	4 4-spinors

Space 7: Octonion

64 dimensions $2^{6/2} = 8$ rows/columns	2 Space-time Dimensions 4 4-spinors (Built from four 16 dimension spaces of the 10 space spectrum)

Figure 2a.5. A sequence of fermion-antifermion annihilations generating instances in the set of Octonion Cosmology spaces.

2a.5 Comments

One might expect that generating the sequence of spaces would take an extraordinarily long time. However time, as we know it, only exists in our space 4 universe. Time is not a problem.

The octonion spaces have many dimensions beyond their space-time dimensions. These additional dimensions furnish the fundamental representations of the internal symmetry groups in the space. We show this for space 4 in chapter 2b in Fig. 2b.2. Fig. 2b.5 shows the *form* of the fermion part of the instance of universe (described by our NEWQUeST theory.) The fermions of our universe have this form.

An interesting feature: The creation of each space instances always leads to a Big Bang —justifying the Big Bang for our universe for, perhaps, the first time.[29]

<u>Point of Clarification</u>: When we say a space (universe) is generated by a fermion-antifermion annihilation we mean that an *instance* of the space is generated. A space is defined independently of its contents. A space plus its contents is called an *instance* of the space. Annihilations generate instances. The source of the matter and energy of an instance is the mass-energy of the annihilating particles. We sometimes say space to refer to a space instance to avoid cumbersome text.

2a.6 The Golden Triangle of Mass, Energy and Space

Examining the Octonion Cosmology spectrum and its origin in fermion-antifermion annihilations, we see a "golden" triangle of transformations between mass, energy, and space. Mass can be transformed into energy and vice versa.

Mass and/or energy can also be transformed into a space and vice versa. The creation of a space (instance) is the result of a mass and energy transformation. A space instance generates mass and energy transformations as it evolves. A space instance also generates instances of particles and of subspaces through fermion-antifermion annihilations within it.

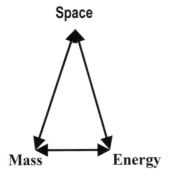

Figure 2a.6. The golden triangle of mass, energy, and space.

---

[29]29 See Blaha (2021d) for the author's model for universe expansion and the Hubble Constant.

A remarkable reflection of the connection between mass and space is the equality of the number of fundamental fermion types in each of the ten spaces and the dimension of its space. In Cayley-Dickson number 4 space, which is that of our universe, we have found 256 fundamental fermion types matching the $16^2 = 256$ dimensions of the space.

2a.7 Octonion Cosmology Connection to Elementary Particle Theory

The Octonion Cosmology with its ten spaces originated in the author's Unified SuperStandard Theory (NEWUST) that was developed over a number of years, and described in:

Blaha, 2018e, *Unification of God Theory and Unified SuperStandard Model THIRD EDITION*

Blaha, 2020a, *Quaternion Unified SuperStandard Theory (The QUeST) and Megaverse Octonion SuperStandard Theory (MOST)*

Blaha, 2020c, *Unified SuperStandard Theories for Quaternion Universes & The Octonion Megaverse*

Blaha, 2021c, *Beyond Octonion Cosmology*

Blaha, 2021d, *Universes are Particles*

Blaha, 2021e, *Octonion-like dna-based life, Universe expansion is decay, Emerging New Physics*

Blaha, 2021f, *The Science of Creation New Quantum Field Theory of Spaces*

and other books by the author.

In the Fall of 2019 the author considered the possibility that the UST was based on a deeper theory and proceeded to consider theories based on hypercomplex numbers such as quaternions and octonions. *Thus Octonion Cosmology did not "emerge out of thin air."* It provided a deeper base for the UST.

Remarkably the author found that the hypercomplex theories, QUeST and NEWQUeST, had similar internal symmetry groups, and fundamental fermions, as UST. In addition a hypercomplex theory NEWUTMOST furnished an acceptable theory of the Megaverse (Multiverse). Over the following two years 2020 and 2021 Octonion Cosmology emerged with its ten spaces.

The history of this effort, which has some relevance to Natural Philosophy, appears below. It is extracted from *Quantum Space Theory With Application to Octonion Cosmology Beyond Octonion Cosmology.*

2b. Octonion Cosmology Connection to the Unified SuperStandard Theory (UST and NEWUST)

The author's UST and NEWUST theories originated in the past twenty years from the Standard Model of Particles with SU(2)⊗U(1)⊗SU(3) internal symmetries combined with Two-Tier Quantum and PseudoQuantum Field Theory (GiFT).

Noting the presence of conserved particle numbers, and the presence of at least three fermion generations, we introduced the U(4) Generation Groups and the U(4) Layer Groups together with four layers of four generation of "Normal" fermions necessitated by Generation Group and Layer group symmetries at the point of the Big Bang prior to symmetry breaking. The Dark sector had a corresponding set of four layers of four generations of "Dark" fermions. The result was the Unified SuperStandard Theory (UST) with the symmetry:

$$\{[SU(2)\otimes U(1)\otimes SU(3)]^2\otimes U(4)^4\}^4$$

supplemented with an additional Strong Interaction U(1) group. Space-time had four dimensions.

In the Fall, 2019 the author discovered that an octonion-based theory that he constructed and named QUeST had the same internal symmetries as UST with the addition of $U(1)^8$. The addition of $U(1)^8$ indicates that the Strong Interactions in the theory are broken Strong SU(4).

QUeST's internal symmetries, which could be based on a 16×16 dimension array of dimensions, was

$$[SU(2)\otimes U(1)\otimes SU(3)\otimes U(1)]^8\otimes U(4)^{16}$$

During 2020 the author developed an octonion space spectrum for both universes and Megaverses, and other spaces. The spectrum was shown to arise from a generation mechanism whereby fermion-antifermion annihilation in a higher space produced an instance of a lower space. A critical part of the derivation of the octonion spectrum was the realization that even space-time dimension spinor arrays are composed of Cayley number rows and columns. Spinor arrays of annihilating fermion-antifermion pairs were shown to generate the arrays of dimensions of subspace instances.

Analyzing the spinor arrays the author noted that the dimension array could be viewed as composed of 64 dimension subblocks, which were further subdivided into 16 dimension subblocks.

The subblock structuring, using the known contents of the Standard Model plus Generation and Layer groups for guidance, gave the dimension array structure containing 4×4 subblocks in Figs. 2b..1 and 2b.2.

Thus there was a *most* satisfactory match between UST and QUeST with the only significant difference being the space-time: four octonion (complex quaternion) coordinates for QUeST and four real space-time coordinates for UST. This difference was resolved in NEWUST and NEWQUeST. See the Connection groups in Fig. 2b.3.

Figs. 2b.4 and 2b.5 show the detailed group and fermion structures.

The form of the square spinor arrays generated by fermion-antifermion annihilation gives 64 dimension blocks and 16 dimension blocks as well as 32 dimension composite blocks that are evidenced in the NEWQUeST fermion spectrums and internal symmetry group structure.

2b.1 Dimensions of Symmetries and Coordinates

Since we see only real dimensions in Reality, we transferred 28 QUeST dimensions from space-time to $U(2)^7$ internal symmetry dimensions. The set of internal symmetries was increased by $U(2)^7$, which we call Connection Groups. Each Connection group specifies interactions between corresponding fermions (e with e, q with q, and so on) in separate layers and between Normal and Dark fermions. The connections between the various blocks of fermions are shown in Fig. 2b.3. *We implement the very practical rule that all blocks must be connected by interactions or they would not be of physical interest. A totally isolated block effectively does not exist physically (except possibly for gravitation effects).*

The interactions of the Connection groups must be very weak and/or their gauge bosons must be very massive.

The addition of the Connection Groups and the reduction of space-time dimensions accordingly results in NEWQUeST and NEWUST as summarized below.

Note: the Generation, Layer, and Connection groups are all badly broken. Their vector bosons must be very massive since they have not been detected in experiments.

2b.1.1 Internal Symmetries

The groups are ElectroWeak $SU(2) \otimes U(1)$, Strong $SU(3)$, Generation Group $U(4)$, Layer Group $U(4)$, and $U(2)$ and $U(4)$ Connection groups obtained by transfer from space-time coordinates (See Blaha 2012c). The $SU(3) \otimes U(1)$ symmetries may be a broken $SU(4)$ symmetries. The internal symmetries for the theories are:

<u>UST</u>

$$[SU(2) \otimes U(1) \otimes SU(3)]^8 \otimes U(4)^{16} \tag{2b.1}$$

<u>QUeST</u>

$$[SU(2) \otimes U(1) \otimes SU(3) \otimes U(1)]^8 \otimes U(4)^{16} \tag{2b.2}$$

<u>NEWQUeST</u>

$$[SU(2) \otimes U(1) \otimes SU(3) \otimes U(1)]^8 \otimes U(4)^{16} \otimes U(2)^7 \tag{2b.3}$$

The only change is in internal symmetries: Twenty-eight real dimensions transferred from space-time coordinates to Connection group $U(2)^7$ internal symmetry.

NEWUST

$$[SU(2) \otimes U(1) \otimes SU(3) \otimes U(1)]^8 \otimes U(4)^{16} \otimes U(2)^7 \qquad (2b.4)$$

The only change in internal symmetries: Twenty-eight real dimensions added for $U(2)^7$ Connection group internal symmetry.

2b.1.2 Space-Time Coordinates

UST

Four real space-time coordinates.

QUeST

Four octonion (complex quaternion) coordinates.

NEWUST

Four real space-time coordinates. No change from UST space-time.

NEWQUeST

Four real space-time coordinates. The six coordinates in the n = 4 octonion space were lowered to four space-time coordinates with two coordinates transferred to Connection groups.

The only change is in space-time coordinates: Fourteen dimensions transferred from QUeST space-time coordinates to Connection group $U(2)^7$ internal symmetry.

2b.2 Fundamental Fermion Spectrum

There are 256 fundamental fermions in NEWQUeST and NEWUST. Conceptually their structure can be viewed as an extrapolation of the known three generations of The Standard Model. For good reason (U(4) Generation groups) a fourth generation was indicated and a corresponding Dark sector of similar structure was added. In addition, because of the need for Layer groups, the overall structure consisted of four copies of this layer (due to U(4) Layer groups).

Correspondingly, each layer also has its own set of internal symmetry gauge groups to limit mixing between the layers to Layer group interactions and Connection group interactions.

Fig. 2b.1 shows the structure of the NEWQUeST/NEWUST fermions. The blocks are 4 × 4 reflecting the origin of the NEWQUeST/NEWUST space (universe) instance from Megaverse fermion-antifermion annihilation. The spinor analysis of their spinor arrays yields a 16 dimension block structure. The 64 dimension fermion layers reflect the 64 dimension structuring of the Megaverse obtained from its creation by fermion-antifermion creation in the Maxiverse.

2b.3 Total Dimensions

The total of internal symmetry and space-time dimensions is 256 in all four theories listed above. It is based on the 16×16 dimension array of the Cayley number n = 4 space of the Octonion spectrum.

2b.4 Pattern of Internal Symmetries

The NEWQUeST dimension array for internal symmetries is subdivided into four layers of 56 dimensions—just as in NEWUST (and UST). Fig. 2b.2 displays the layers using SU(4) in place of SU(3)⊗U(1). Each layer has a block of 28 dimensions for Normal and 28 dimensions for Dark sectors. There are also seven U(2) Connection groups plus four real-valued space-time coordinates bringing the NEWQUeST total to 256 dimensions.= 4*56 + 28 + 4 = 256 dimensions. The Connection groups are shown in Fig. 2b.3.

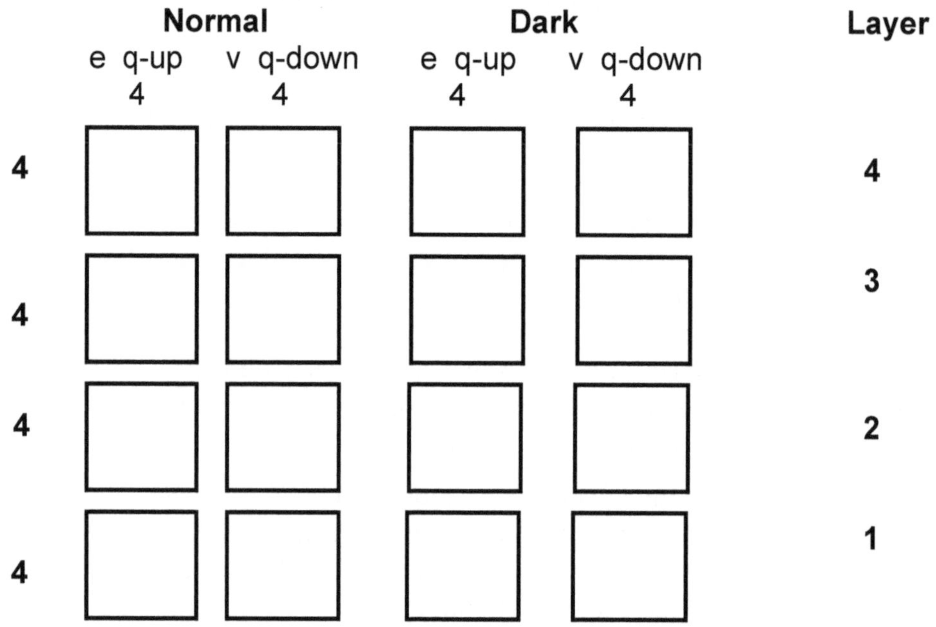

Figure 2b.1. Block form of a 16×16 NEWQUeST/NEWUST fermion array with each block row corresponding to one layer. Each block contains four generations of fermions. The result is 4×4 blocks. The label e q-up indicates a charged lepton – up-type quark pair, v q-down indicates a neutral lepton – down-type quark pair, and so on. The blocks can be viewed as SU(3)⊗U(1) or broken SU(4) blocks.

Layers	NORMAL		DARK	
↓	4	4	4	4
4	SU(2)⊗U(1)⊗SU(3)⊗U(1) 4 Space-time Dimensions	Generation + Layer Groups	SU(2)⊗U(1)⊗SU(3)⊗U(1) 4 Space-time Dimensions	Generation + Layer Groups
4	SU(2)⊗U(1)⊗SU(3)⊗U(1) 4 Space-time Dimensions	Generation + Layer Groups	SU(2)⊗U(1)⊗SU(3)⊗U(1) 4 Space-time Dimensions	Generation + Layer Groups
4	SU(2)⊗U(1)⊗SU(3)⊗U(1) 4 Space-time Dimensions	Generation + Layer Groups	SU(2)⊗U(1)⊗SU(3)⊗U(1) 4 Space-time Dimensions	Generation + Layer Groups
4	SU(2)⊗U(1)⊗SU(3)⊗U(1) 4 Space-time Dimensions	Generation + Layer Groups	SU(2)⊗U(1)⊗SU(3)⊗U(1) 4 Space-time Dimensions	Generation + Layer Groups

Figure 2b.2.. Four layers of Internal Symmetry groups in NEWQUeST and NEWUST (omitting Connection Groups) showing the 4 by 4 subblocks. The groups in each layer are independent of those in other layers. The groups in each subblock of each layer are independent of those in the other subblocks. Each subblock contains 16 dimensions. The block dimensions furnish fundamental representations for the groups listed. The entire set of blocks contains 256 dimensions. Each layer contains 56 internal symmetry dimensions. The first two columns are for the "Normal" sector. The last two columns are for the "Dark" sector (although most of the Normal sector is Dark observationally at present.) This figure also holds for UST with the addition of U(1) groups. The eight sets of 4 real dimension space-times combine to give a 4 real dimension space-time and seven U(2) Connection groups.

Connection Group Applied to Fermions

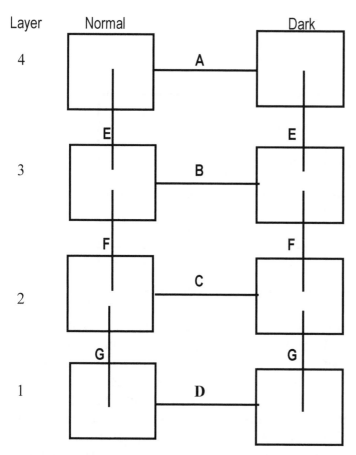

Figure 2b.3. The seven U(2) Connection groups (shown as 10 lines) between the eight NEWQUeST/NEWUST blocks. Connection groups are obtained by transfering 28 dimensions from QUeST space-time to internal symmetries with the consequent reduction of the space-time from four octonion (complex quaternion) coordinates to four real coordinates. The Connection groups generate rotations and interactions between corresponding fermions and vector bosons of each pair of blocks. The Normal and Dark sector U(2) vertical connections above (E, F, G) represent the same U(2) groups.

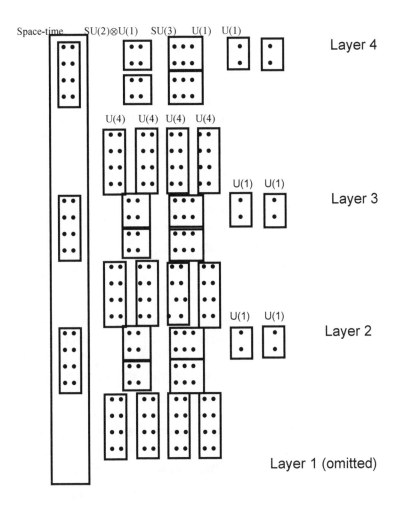

Figure 2b.4. Three of the four layers of QUeST internal symmetry groups (and space-time) for Cayley-Dickson space 4. Layer 1 which has an identical form was omitted due to "page space" limitations. Note the left column of blocks combine to specify a 4 dimension octonion space-time. Note each layer has 64 dimensions.

The Fermion Periodic Table

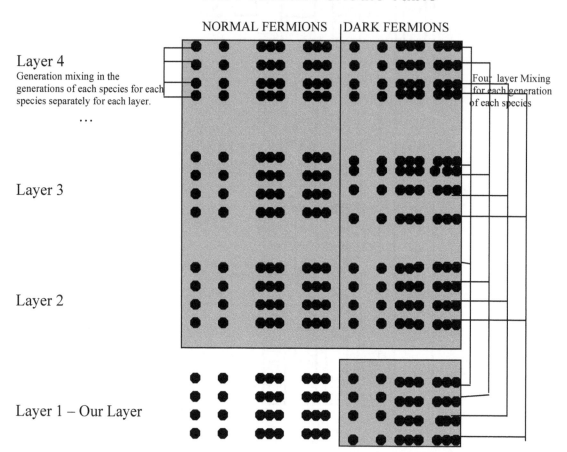

Figure 2b.5. Fermion particle spectrum and partial examples of the pattern of mass mixing of the Generation group and of the Layer group. Unshaded parts are the known fermions with an additional, as yet not found, 4th generation. The lines on the left side (only shown for one layer) display the Generation mixing within each layer. The Generation mixing occurs within each layer using a separate Generation group for each layer. The lines on the right side show Layer group mixing (for Dark matter) with the mixing among all four layers for each of the four generations individually. There are four Layer groups for Normal matter and four Layer groups for Dark matter.. There are 256 fundamental fermions. QUeST and UST have the same fermion spectrum.

3. GiFT for *Scalar* Particle Creation/Annihilation Operator Transformations

In Blaha (2021i) we developed the features of Generalized Field Theory (GiFT). In this chapter we apply the formalism to the case of fermions. We show spin is naturally incorporated within GiFT. We begin by considering two 4-dimension coordinate systems denoted α and β and develop a relation between the quantizations in these coordinate systems in the case where the spins of the fermions in the respective coordinate systems are the same. Then we consider the case where the spins of states in one coordinate system correspond to superpositions of spin states in the other coordinate system.

Since higher dimension space-times appear in Octonion Cosmology and other theories we develop the relation between quantizations of coordinate systems in higher dimensions. We then consider the relations of these results to other spin-based theories such as Twistor Theory.

3.1 GiFT for *Scalar* Creation/Annihilation Operator Transformations

We illustrate the procedure by considering the case of a scalar particle in four dimensions[30] and expand on the development in section 4 of S. Blaha, "The Local Definition of Asymptotic Particle States", IL Nuovo Cimento **49A**, 35 (1979), which we call I. We associate the scalar particle with two scalar fields: $\varphi_1(x)$ and $\varphi_2(x)$. We choose $\varphi_1(x)$ to have a zero equal time commutator with $d\varphi_1(x)/dx^0$ and $\varphi_2(x)$ to have a conventional equal time commutator with $d\varphi_2(x)/dx^0$. Conceptually $\varphi_1(x)$ is a "classical" field and $\varphi_2(x)$ is a quantum field. A Lagrangian that implements these choices of commutation relations is:

$$\mathcal{L} = \partial^\mu \varphi_1(x)\partial_\mu\varphi_2(x) - \tfrac{1}{2}\,\partial^\mu \varphi_1(x)\partial_\mu\varphi_1(x) - m^2\,\varphi_1(x)\varphi_2(x) + \tfrac{1}{2}\,m^2\,\varphi_1(x)^2 \qquad (3.1)$$

$$(\Box + m^2)\varphi_1(x) = 0 \qquad (3.2)$$

$$(\Box + m^2)\varphi_2(x) - (\Box + m^2)\varphi_1(x) \quad = 0 \qquad (3.3)$$

The canonical momenta are

$$\pi_1 = d\varphi_2(x)/dt - d\varphi_1(x)/dt \qquad (3.4)$$
$$\pi_2 = d\varphi_1(x)/dt \qquad (3.5)$$

and the equal time commutation relations are

---

[30] The case of a space with a higher dimension spacetime is completely analogous.

$$[\varphi_i(x), \pi_j(y)] = i\delta_{ij}\delta(\mathbf{x} - \mathbf{y}) \tag{3.6}$$

$$[\varphi_i(x), \varphi_j(y)] = [\pi_i(x), \pi_j(y)] = 0 \tag{3.7}$$

for i, j = 1, 2 implying

$$[\varphi_1(x), d\varphi_1(y)/dt] = 0 \tag{3.8}$$

$$[\varphi_2(x), d\varphi_2(y)/dt] = i\delta^3(\mathbf{x} - \mathbf{y}) \tag{3.9}$$

$$[\varphi_1(x), d\varphi_2(y)/dt] = i\delta^3(\mathbf{x} - \mathbf{y}) \tag{3.10}$$

The *most* general mode expansion of the fields in the coordinate system A with Fourier momentum α is

$$\varphi_{1A}(x) = \Sigma_\alpha \, [a_{1\alpha} \, f_\alpha(x) + a^\dagger_{1\alpha} \, f_\alpha^*(x)] \tag{3.11}$$

$$\varphi_{2A}(x) = \Sigma_\alpha \, [a_{2\alpha} f_\alpha(x) + a^\dagger_{2\alpha} \, f_\alpha^*(x)] \tag{3.12}$$

and in coordinate system B with Fourier momentum β is

$$\varphi_{1B}(x) = \Sigma_\beta \, [a_{1\beta} \, f_\beta(x) + a^\dagger_{1\beta} \, f_\beta^*(x)] \tag{3.13}$$

$$\varphi_{2B}(x) = \Sigma_\beta \, [a_{2\beta} \, f_\beta(x) + a^\dagger_{2\beta} \, f_\beta^*(x)] \tag{3.14}$$

The creation/annihilation operators of A have the commutation relations:

$$[a_{i\alpha'}, a_\alpha] = 0 \tag{3.15}$$
$$[a^\dagger_{i\alpha'}, a^\dagger_{j\alpha}] = 0$$

$$[a_{i\alpha}, a^\dagger_{j\alpha'}] = (1 - \delta_{ij})\delta^3(\boldsymbol{\alpha} - \boldsymbol{\alpha'}) \tag{3.16}$$

with i, j = 1, 2. Two related vacuums are defined: |0>₁ and |0>₂ by

$$a_{1\alpha}|0>_{2\alpha} = a^\dagger_{1\alpha}|0>_{2\alpha} = 0 \tag{3.17}$$
$$a_{2\alpha}|0>_{1\alpha} = a^\dagger_{2\alpha}|0>_{1\alpha} = 0$$

and

$$a_{i\alpha}|0>_{i\alpha} \neq 0 \qquad a^\dagger_{i\alpha}|0>_{i\alpha} \neq 0 \tag{3.18}$$

for i = 1, 2 and all α and α'. And similarly for the creation/annihilation operators of B with α replaced with β.

The general coordinate system B Fourier expansion in terms of A creation/annihilation operators is

$$\varphi_{1B}(x) = \Sigma_\beta \, \Sigma_\alpha \, [(c_{11}a_{1\alpha} + c_{12}a_{2\alpha} + C_{11}a^\dagger_{1\alpha} + C_{12}a^\dagger_{2\alpha})f_\beta(x) + \\ + (c'_{11}a^\dagger_{1\alpha} + c'_{12}a^\dagger_{2\alpha} + C'_{11}a_{1\alpha} + C'_{12}a_{2\alpha})f_\beta^*(x)] \tag{3.19}$$

$$\varphi_{2B}(x) = \Sigma_\beta \, \Sigma_\alpha \, [(c_{21}a_{1\alpha} + c_{22}a_{2\alpha} + C_{21}a^\dagger_{1\alpha} + C_{22}a^\dagger_{2\alpha}) \, f_\beta(x) + \\ + (c'_{21}a^\dagger_{1\alpha} + c'_{22}a^\dagger_{2\alpha} + C'_{21}a_{1\alpha} + C'_{22}a_{2\alpha})f_\beta^*(x)] \tag{3.20}$$

where the c_{ij} and C_{ij}, and c'_{ij} and C'_{ij} are all functions of α.

We define a sets of Bogoliubov transformations B_1 and B_2 that give

$$a_{i\beta} = \Sigma_\alpha \, (f_\beta, f_\alpha) B_{2\alpha} B_{1\alpha} a_{i\alpha} (B_{1\alpha} B_{2\alpha})^{-1} = \Sigma_\alpha \, (f_\beta, f_\alpha)[c_{i1}a_{1\alpha} + c_{i2}a_{2\alpha} + C_{i1}a^\dagger_{1\alpha} + C_{i2}a^\dagger_{2\alpha}]$$
(3.21)

$$a^\dagger_{i\beta} = \Sigma_\alpha \, (f^*_\beta, f^*_\alpha) \, B_{2\alpha} B_{1\alpha} a^\dagger_{i\alpha} (B_{1\alpha} B_{2\alpha})^{-1} = \Sigma_\alpha \, (f^*_\beta, f^*_\alpha)[c'_{i1}a^\dagger_{1\alpha} + c'_{i2}a^\dagger_{2\alpha} + C'_{i1}a_{1\alpha} + C'_{i2}a_{2\alpha}]$$
(3.22)

3.1.1 The B_1 Transformation

The B_1 transformation has the form:

$$B_{1\alpha}(x_1, x_2) = \exp[x_1\Gamma^1_{3\alpha}] \exp[x_2\Gamma^1_{2\,\alpha}]$$
(3.23)

where the Hermitian operators $\Gamma^1_{1i\alpha}$ are

$$\Gamma^1_{3\alpha} = (a^\dagger_{2\alpha}a_{1\alpha} + a_{2\alpha}a^\dagger_{1\alpha})/2$$
(3.24)
$$\Gamma^1_{2\alpha} = i(a^\dagger_{2\alpha}a^\dagger_{1\alpha} - a_{2\alpha}a_{1\alpha})/2$$
(3.25)

We also define

$$\Gamma^1_{1\alpha} = -(a^\dagger_{2\alpha}a^\dagger_{1\alpha} + a_{2\alpha}a_{1\alpha})/2$$
(3.26)

and

$$\Gamma^1_{4\alpha} = (a^\dagger_{2\alpha}a_{1\alpha} - a_{2\alpha}a^\dagger_{1\alpha})/2$$
(3.27)

Together the three operators satisfy su(1,1) algebra commutation relations:

$$[\Gamma^1_{1\alpha}, \Gamma^1_{2\alpha'}] = -i\delta_{\alpha\alpha'} \, \Gamma^1_{3\alpha} \qquad [\Gamma^1_{2\alpha}, \Gamma^1_{3\alpha'}] = i\delta_{\alpha\alpha'} \, \Gamma^1_{1\alpha} \qquad [\Gamma^1_{3\alpha}, \Gamma^1_{1\alpha'}] = i\delta_{\alpha\alpha'} \, \Gamma^1_{2\alpha}$$
(3.28)

They transform $a_{i\alpha}$ and $a^\dagger_{i\alpha}$ to terms of the form $c_{i\alpha}a_{i\alpha} + c_{i\alpha}'a^\dagger_{i\alpha}$ where $c_{i\alpha}$ and $c_{i\alpha}'$ are constants.

Some properties of the su(1,1) algebra are detailed in S. Blaha, "The Local Definition of Asymptotic Particle States", IL Nuovo Cimento **49A**, 35 (1979).

3.1.1 The B_2 Transformation

The B_2 transformation has the form:

$$B_{2\alpha}(x_1, x_2) = \exp[x_1\Gamma^2_{3\alpha}] \exp[x_2\Gamma^2_{2\,\alpha}]$$
(3.29)

where the Hermitian operators $\Gamma^2_{1i\alpha}$ are

$$\Gamma^2_{3\alpha} = (a^\dagger_{1\alpha}a_{1\alpha} + a_{1\alpha}a^\dagger_{1\alpha})/2$$
(3.30)
$$\Gamma^2_{2\alpha} = (a^\dagger_{2\alpha}a_{2\alpha} + a_{2\alpha}a^\dagger_{2\alpha})/2$$
(3.30)
$$\Gamma^2_{1\alpha} = (a^\dagger_{2\alpha}a^\dagger_{2\alpha} + a_{2\alpha}a_{2\alpha})/2$$
(3.31)

We also define the Hermitian operators

$$\Gamma^2_{4\alpha} = (a^\dagger_{1\alpha} a^\dagger_{1\alpha} + a_{1\alpha} a_{1\alpha})/2 \tag{3.32}$$

and

$$\Gamma^2_{5\alpha} = i(a^\dagger_{2\alpha} a^\dagger_{2\alpha} - a_{2\alpha} a_{2\alpha})/2 \tag{3.33}$$

$$\Gamma^2_{6\alpha} = i(a^\dagger_{1\alpha} a^\dagger_{1\alpha} - a_{1\alpha} a_{1\alpha})/2 \tag{3.34}$$

They transform $a_{1\alpha}$ and $a^\dagger_{1\alpha}$ to have added terms of the form $c_\alpha a_{2\alpha} + c_\alpha' a^\dagger_{2\alpha}$ where c_α and c_α' are constants; and they transform $a_{2\alpha}$ and $a^\dagger_{2\alpha}$ to have added terms of the form $c_\alpha a_{1\alpha} + c_\alpha' a^\dagger_{1\alpha}$.

The complete set of Γ operators are part of an su(1,1) group. This group appears in Twistor Theory. A relation/unification of the group of these Bogoliubov transformations and Twistor Theory may eventually emerge.

The transformation $B_{2\alpha} B_{1\alpha}$ generates the maps of eqs. 3.21 and 3.22. As we will see next the key terms for the transition between annihilation/creation operators of two coordinate systems, A and B, are in

$$[c'_{i1} a^\dagger_{1\alpha} + c'_{i2} a^\dagger_{2\alpha} + C'_{i1} a_{1\alpha} + C'_{i2} a_{2\alpha}] \tag{3.35}$$

if we construct states based on $a^\dagger_{2\beta}$ and $a^\dagger_{2\alpha}$ with the vacuum $|0\rangle_2$. Note that there are four constants, which are functions of α, in eq. 3.35. These constants are determined by the four parameters in $B_{2\alpha} B_{1\alpha}$. Thus the transformation between coordinate systems embodied in $B_{2\alpha} B_{1\alpha}$ determines the constants in eq. 3.35.

3.1.2 States in the Respective Coordinate Systems

The B states are superpositions of the A states as eq. 3.21 and 3.22 demonstrate. The one particle state in the B coordinate system is

$$a^\dagger_{2\beta} |0\rangle_2 = \Sigma_\alpha (f^*_\beta, f^*_\alpha)[c'_{21} a^\dagger_{1\alpha} + c'_{22} a^\dagger_{2\alpha} + C'_{21} a_{1\alpha} + C'_{22} a_{2\alpha}] |0\rangle_2$$

$$= \Sigma_\alpha (f^*_\beta, f^*_\alpha)[c'_{22} a^\dagger_{2\alpha} + C'_{22} a_{2\alpha}] |0\rangle_2 \tag{3.36}$$

where the $a^\dagger_{2\alpha}$ term generates a one particle a state, and the $a_{2\alpha}$ term generates an A particle negative energy state. The two particle B coordinate system state

$$a^\dagger_{2\beta} a^\dagger_{2\beta} |0\rangle_2$$

generates a superposition in momenta α) of two A particles, two A negative energy particles and an A vacuum state term.

Later, when we consider fermions, we will see that parton momentum distributions, $xF(x)$, in deep inelastic e-p scattering may partially originate in the transition from a rest frame coordinate system to a different internal coordinate system.

3.2 Scalar Particle Creation/Annihilation Space (CASe)

The set of creation and annihilation operators a_1, a_2, $a^\dagger_1$, $a^\dagger_2$ form an operator basis for spaces, for each value α. In the scalar particle case in four space-time dimensions we define a four dimension space that we call (1,1) Creation/Annihilation Space or CASe(1,1) that supports an su(1,1) representation for each α, which we call a GiFT transformation. Defining basis "ortho" vectors $a_\alpha = (a^\dagger_{1\alpha}, a^\dagger_{2\alpha}, a_{1\alpha}, a_{2\alpha})$ we specify complex coordinates for an irreducible su(1,1):

$$x_\alpha = (x_{1\alpha}{}^0, x_{2\alpha}{}^0, x_\alpha{}^1, x_{2\alpha}{}^1) \qquad (3.37)$$

Defining the metric

$$ds^2 = |dx_{1\alpha}{}^0|^2 + |dx_{2\alpha}{}^0|^2 - |dx_{1\alpha}{}^1|^2 - |dx_{2\alpha}{}^1|^2 \qquad (3.38)$$

we note that there is an su(1,1) group invariance.

Thus for scalar particles an su(1,1) symmetry exists for each Fourier component α in a space-time of any number of space-time dimensions.

The vectors in the space have the form

$$a_\alpha \cdot x_\alpha = x_{1\alpha}{}^0 a^\dagger_{1\alpha} + x_{2\alpha}{}^0 a^\dagger_{2\alpha} + x_\alpha{}^1 a_{1\alpha} + x_{2\alpha}{}^1 a_{2\alpha} \qquad (3.39)$$

The "ortho" vectors a_α form an algebra reminiscent of hypercomplex quaternions. The lack of complete commutativity is not a source for disquiet when one considers the anti-commutativity of quaternion, octonion and sedenion multiplication. It can be viewed positively as a step in the direction of a larger unified theory such as

$$\text{Spinor Space+CASe(n,m)} \qquad (3.40)$$

CASe is enlarged for fermions in the various space-time dimensions. See chapter 4.

Note: A General Relativistic transformation between coordinate systems with the same Killing vector does not induce a GiFT transformation. A General Relativistic transformation between coordinate systems with differing sets of Killing vectors (or to a coordinate system with no Killing vector) does induce a GiFT CASe transformation.

4. GiFT for *Fermion* Creation/Annihilation Operator Transformations

The GiFT formalism developed in Blaha (2021i) and in section 4 of S. Blaha, "The Local Definition of Asymptotic Particle States", IL Nuovo Cimento **49A**, 35 (1979), which we call I, is directly extendable to fermions. We begin by considering 4-dimension space-time coordinate systems. Following the framework developed in section 4 of I we define fermion fields $\psi_1(x)$ and $\psi_2(x)$ in a coordinate system A labeled with α as

$$\psi_{1A}(x) = \Sigma_{\alpha,s}[b_{1\alpha s}u_{\alpha s}f_\alpha(x) + d^\dagger_{1\alpha s}v_{\alpha s}f_\alpha{}^*(x)] \tag{4.1}$$

$$\psi_{2A}(x) = \Sigma_{\alpha,s}[b_{2\alpha s}u_{\alpha s}f_\alpha(x) + d^\dagger_{2\alpha s}v_{\alpha s}f_\alpha{}^*(x)] \tag{4.2}$$

and in another coordinate system B labeled with β as

$$\psi_{1B}(x) = \Sigma_{\beta,s}[b_{1\beta s}u_{\beta s}\, g_\beta(x) + d^\dagger_{1\beta s}v_{\beta s}\, g_\beta{}^*(x)] \tag{4.3}$$

$$\psi_{2B}(x) = \Sigma_{\beta,s}[b_{2\beta s}u_{\beta s}\, g_\beta(x) + d^\dagger_{2\beta s}v_{\beta s}\, g_\beta{}^*(x)] \tag{4.4}$$

where $f_\alpha(x)$ and $g_\beta(x)$ are Fourier components.

To begin we assume the spin sectors transform independently We also anticipate the Fourier expansions are related in a manner similar to the scalar case of chapter 3. Then for spin value s we have

$$b_{1\beta s} = \Sigma_{\alpha,x}\, (g_\beta, f_\alpha)\, u^\dagger{}_{\beta s}u_{\alpha s}\, (c_{11s}b_{1\alpha s} + c_{12s}b_{2\alpha s} + C_{11s}b^\dagger{}_{1\alpha s} + C_{12s}b^\dagger{}_{2\alpha s}) \tag{4.6a}$$

$$b_{2\beta s} = \Sigma_{\alpha,x}\, (g_\beta, f_\alpha)\, u^\dagger{}_{\beta s}u_{\alpha s}\, (c_{21s}b_{1\alpha s} + c_{22s}\, b_{2\alpha s} + C_{21s}b^\dagger{}_{1\alpha s} + C_{22s}b^\dagger{}_{2\alpha s}) \tag{4.6b}$$

$$d^\dagger{}_{1\beta s} = \Sigma_{\alpha,x}\, (g^*{}_\beta\, f^*{}_\alpha)\, v^\dagger{}_{\beta s}v_{\alpha s}\, (c'_{11s}d^\dagger{}_{1\alpha s} + c'_{12s}d^\dagger{}_{2\alpha s} + C'_{11s}d^\dagger{}_{1\alpha s} + C'_{12s}d^\dagger{}_{2\alpha s}) \tag{4.6c}$$

$$d^\dagger{}_{2\beta s} = \Sigma_{\alpha,x}\, (g^*{}_\beta\, f^*{}_\alpha)\, v^\dagger{}_{\beta s}v_{\alpha s}\, (c'_{21s}d^\dagger{}_{1\alpha s} + c'_{22s}d^\dagger{}_{2\alpha s} + C'_{21s}d^\dagger{}_{1\alpha s} + C'_{22s}d^\dagger{}_{2\alpha s}) \tag{4.6d}$$

generalizing eq. 118 of I where the c_{ij} are all constants and where x is a space-like surface. The inner products have the form

$$\Sigma_x\, (g_\beta, f_\alpha) = i\, \Sigma_x\, g^*{}_\beta \overset{\leftrightarrow}{\partial_0} f_\alpha \tag{4.6e}$$

where "0" indicates the time-like coordinate. For example, in rectilinear coordinates

$$\Sigma_x \, (g_{k'}, f_k) = i \, \Sigma_x \, g^*_k \overset{\leftrightarrow}{\partial}_0 f_k = \delta^3(\mathbf{k'} - \mathbf{k}) \qquad (4.6e)$$

for momenta k and k'.

The non-zero anti-commutation relations of the creation/annihilation operators (with those of the rest being zero) are:

$$\{b_{i\alpha s}(\alpha), \, b^\dagger_{j\alpha s'}(\alpha')\} = \{d_{i\alpha s}(\alpha), \, d^\dagger_{j\alpha s'}(\alpha')\} = (1 - \delta_{ij})\delta_{ss'} \, \delta^3(\boldsymbol{\alpha} - \boldsymbol{\alpha}') \qquad (4.7)$$
$$\{b_{i\beta s}(\beta), \, b^\dagger_{j\beta s'}(\beta')\} = \{d_{i\beta s}(\beta), \, d^\dagger_{j\beta s'}(\beta')\} = (1 - \delta_{ij})\delta_{ss'} \, \delta^3(\boldsymbol{\beta} - \boldsymbol{\beta}')$$

Note the transformations of eq. 4.6 superimpose "momenta" in relating the coordinate system quantizations. But they do not superimpose spins.[31] As a result the transformations can be factored into two sets: one set for spin up and one set for spin down.

4.1 Transformations of the b and b† Operators

The sets of Bogoliubov transformations B_1 and B_2 for spin up or down b creation/annihilation operators are defined by

$$B_{1\alpha}(x_1, x_2) = ex[x_1\Gamma^1{}_{3\alpha}] \exp[x_2\Gamma^1{}_{2\,\alpha}] \qquad (4.8)$$

where the Hermitian operators $\Gamma^1{}_{1i\alpha}$ are

$$\Gamma^1{}_{3\alpha}(b_{1\alpha s}, b_{2\alpha s}) = (b^\dagger{}_{2\alpha s}b_{1\alpha s} + b_{2\alpha s}b^\dagger{}_{1\alpha s})/2 \qquad (4.9)$$
$$\Gamma^1{}_{2\alpha}(b_{1\alpha s}, b_{2\alpha s}) = i(b^\dagger{}_{2\alpha s}b^\dagger{}_{1\alpha s} - b_{2\alpha s}b_{1\alpha s})/2 \qquad (4.10)$$

We also define

$$\Gamma^1{}_{1\alpha}(b_{1\alpha s}, b_{2\alpha s}) = -(b^\dagger{}_{2\alpha s}b^\dagger{}_{1\alpha s} + b_{2\alpha s}b_{1\alpha s})/2 \qquad (4.11)$$

and

$$\Gamma^1{}_{4\alpha}(b_{1\alpha s}, b_{2\alpha s}) = -(b^\dagger{}_{2\alpha s}b_{1\alpha s} + b_{2\alpha s}b^\dagger{}_{1\alpha s})/2 \qquad (4.12)$$

The B_2 transformation has the form:

$$B_{2\alpha}(x_1, x_2) = \exp[x_1\Gamma^2{}_{3\alpha}] \exp[x_2\Gamma^2{}_{2\,\alpha}] \qquad (4.13)$$

where the Hermitian operators $\Gamma^2{}_{1i\alpha}$ are

$$\Gamma^2{}_{3\alpha} = \Gamma^1{}_{3\alpha}(b_{1\alpha s}, b_{1\alpha s}) \qquad (4.14)$$
$$\Gamma^2{}_{2\alpha} = \Gamma^1{}_{3\alpha}(b_{2\alpha s}, b_{2\alpha s}) \qquad (4.15)$$
$$\Gamma^2{}_{1\alpha} = \Gamma^1{}_{1\alpha}(b_{2\alpha s}, b_{2\alpha s}) \qquad (4.16)$$

We also define the Hermitian operators

---

[31] We consider spin mixing later.

$$\Gamma^2{}_{4\alpha} = \Gamma^1{}_{1\alpha}(b_{1\alpha s}, b_{1\alpha s}) \qquad (4.17)$$
$$\Gamma^2{}_{5\alpha} = \Gamma^1{}_{2\alpha}(b_{2\alpha s}, b_{2\alpha s}) \qquad (4.18)$$
$$\Gamma^2{}_{6\alpha} = \Gamma^1{}_{2\alpha}(b_{1\alpha s}, b_{1\alpha s}) \qquad (4.19)$$

The above set of operators generate the su(1,1) transformations to the forms of eqs. 4.6a and 4.6b. The su(1,1) representation has four real components for b_i, and $b_i^\dagger$ for $i = 1, 2$.

4.2 Transformations of the d and d† Operators

These operators for "holes" have an analogous form to the b operators. They have the same form as those in eqs. 4.8 – 4.19 with each "b" replaced with a "d". These operators generate the transformations to the forms of eqs. 4.6c and 4.6d. These operators are part of the operators of another su(1, 1) in a manner similar to those of chapter 3.

These operators, plus those of section 4.1, are part of the operators of an su(2, 2), which contains operators mixing b and d operators. The su(2,2) representation has eight real components for b_i, and $b_i^\dagger$ and for d_i, and $d_i^\dagger$ for $i = 1, 2$ for each spin.

$$su(1,1) \otimes su(1,1) \rightarrow su(2,2) \qquad (4.20)$$

4.3 Operators of Types b and d *with* Spin Mixing

We can anticipate that certain General Relativistic transformations may cause spin mixing. Some typical (2,2) Creation/Annihilation Space (CASe) transformations are (in the notation of section 4.1)

$$B_{1\alpha}(x_1, x_2) = \exp[x_1 \Gamma^1{}_{3\alpha}] \exp[x_2 \Gamma^1{}_{2\alpha}] \qquad (4.21)$$

with

$$\Gamma^1{}_{3\alpha}(b_{1\alpha s}, b_{2\alpha s'})$$
$$\Gamma^1{}_{2\alpha}(b_{1\alpha s}, b_{2\alpha s'})$$
$$\Gamma^1{}_{1\alpha}(b_{1\alpha s}, b_{2\alpha s'})$$
$$\Gamma^1{}_{4\alpha}(b_{1\alpha s}, b_{2\alpha s'})$$

where $s \neq s'$.

The B_2 transformation has the form:

$$B_{2\alpha}(x_1, x_2) = \exp[x_1 \Gamma^2{}_{3\alpha}] \exp[x_2 \Gamma^2{}_{2\alpha}] \qquad (4.22)$$

where

$$\Gamma^2{}_{3\alpha} = \Gamma^1{}_{3\alpha}(b_{1\alpha s}, b_{1\alpha s'})$$
$$\Gamma^2{}_{2\alpha} = \Gamma^1{}_{3\alpha}(b_{2\alpha s}, b_{2\alpha s'})$$
$$\Gamma^2{}_{1\alpha} = \Gamma^1{}_{1\alpha}(b_{2\alpha s}, b_{2\alpha s'})$$
$$\Gamma^2{}_{4\alpha} = \Gamma^1{}_{1\alpha}(b_{1\alpha s}, b_{1\alpha s'})$$
$$\Gamma^2{}_{5\alpha} = \Gamma^1{}_{2\alpha}(b_{2\alpha s}, b_{2\alpha s'})$$
$$\Gamma^2{}_{6\alpha} = \Gamma^1{}_{2\alpha}(b_{1\alpha s}, b_{1\alpha s'})$$

and $s \neq s'$.

4.3.1 Four Dimension Case

In four dimensions there are two spin values, up and down. Consequently the combination of spin mixed transformations gives su(4,4) as the CASe group, which contains the combined spin up and spin down groups. The su(4,4) representation has sixteen real components for b_i, and $b_i^\dagger$, and d_i, and $d_i^\dagger$, for $i = 1, 2$ for both spins.

$$su(4,4) \tag{4.23}$$

4.3.2 Higher Dimension Case

For space-time dimension r there are $2^{r/2-1}$ spin values. In this case there is an $su(2^{r/2}, 2^{r/2})$ CASe group containing the direct product subgroup:

$$\bigotimes_{k=1}^{2^{r/2-1}} su(2,2)_k \tag{4.24}$$

where k labels the spin value. The combined spin mixing gives the CASe group

$$su(2^{r/2}, 2^{r/2}) \tag{4.25}$$

with $2^{r/2+2}$ real components for b_i, and $b_i^\dagger$, and d_i, and $d_i^\dagger$, for $i = 1, 2$ for all spins.

4.4 Operators of Types b and d *with* Spin and b – d Mixing

If we consider transformations with spin and b – d mixing then the form of the transformed creation/annihilation eq. 4.6a operator becomes

$$b_{1\beta s} = \Sigma_{\alpha,x,s} (g_\beta, f_\alpha) u^\dagger_{\beta s} u_{\alpha s} (c_{11s}b_{1\alpha s} + c_{12s}b_{2\alpha s} + C_{11s}b^\dagger_{1\alpha s} + C_{12}b^\dagger_{2\alpha s} + \tag{4.26}$$
$$+ c'_{11s}d_{1\alpha s} + c'_{12s}d_{2\alpha s} + C'_{11s}d^\dagger_{1\alpha} + C'_{12s}d^\dagger_{2\alpha})$$

with similar forms for the other transformed operators.

The transformations of this type form require an extension of su(4, 4) to su(4, 4) over sedenion hypercharge numbers. (Sedenions have 16 components. They are the next Cayley-Dickson number after octonions.)

In the present case we therefore express each component creation/annihilation operator in eqs. 4.1 and 4.2 as sedenions, and then perform transformations with sedenion array multiplication to generate the terms of the form of eq. 4.26.

4.5 The CASe Groups

The CASe groups of fermions of the universes of Octonion Cosmology are determined above in eq. 4.25. Fig. 4.1 presents them for b and d creation/annihilation operators separately. They have spin mixing and generate transformations of the type of eq. 4.6.

A General Relativistic transformation to a non-static coordinate system without a Killing vector causes fermion wave function creation/annihilation operators to be CASe transformed in every Octonion Cosmology universe (space-time). The correspondence between General Relativistic transformations and the CASe transformations of GiFT reflects the unity of General Relativity and Quantum Theory.

CASe Groups of Eight of the Ten Octonion Universes

Octonion Space Number O_S	Cayley-Dickson Number n	Cayley Number d_c	Dimension Array Total d_d	Space-time-Dimension r	Fermion Spinor Array Total d_s	CASe Group $su(2^{r/2}, 2^{r/2})$ CASe
0	10	1024	1024×1024	18	512×512	$su(512,512)$
1	9	512	512×512	16	256×256	$su(256,256)$
2	8	256	256×256	14	128×128	$su(128,128)$
3	7	128	128×128	12	64×64	$su(64,64)$
4	6	64	64×64	10	32×32	$su(32,32)$
5	5	32	32×32	8	16×16	$su(16,16)$
6	**4**	**16**	$\mathbf{16 \times 16}$	$\mathbf{6 \rightarrow 4}$	$\mathbf{8 \times 8}$	$su(4,4)$
7	3	8	8×8	4	4×4	$su(4,4)$

Figure 4.1. The CASe groups of the eight Octonion Cosmology spaces. Space n = 4 has 6 – 2 = 4 space-time dimensions due to compacted dimensions.

5. A General Relativistic Basis for Octonion Cosmology

The spaces of Octonion Cosmology have been derived from a recursive generation of subspaces through fermion-antifermion annihilation of space fermion particles. In this chapter we consider an alternate, not necessarily unrelated, approach based on General Relativistic considerations. We begin by considering a universe with a 4-dimension space-time. Then we go on to consider the case of spaces with higher dimension space-times in chapters 6 and 7.

5.1 Universe with Four Dimension Space-Time

We begin by considering a space with a four dimension space-time[32] such as our universe. In that space we can define a GiFT formulation for fermions as we did in chapter 4. The fermion wave function has an su(4,4) (CASe) group for transformations of the b and d creation/annihilation operators from a coordinate system to a non-static coordinate system without a Killing vector. These transformations parallel the transformations of General Relativity (GR).

It is worth noting that almost all coordinate systems are non-static.[33] Consequently the set of GR transformations is dominated by transformations to non-static coordinate systems. A CASe transformation to a non-static coordinate system has the effect of changing the Fermi sea such that some fermions are raised above or below the sea. Thus GR transformations cause the set of physical fermion states to be modified resulting in fermion creation/annihilation.

5.1.1 Matrix Representations of su(1,1)

The matrix representations of the su(4,4) group form a 16×16 fundamental representation in a 16 real dimension space. The space vectors have a form analogous to the set of creation/annihilation operators:

$$(b_{1\uparrow}, b_{2\uparrow}, b^{\dagger}_{1\uparrow}, b^{\dagger}_{2\uparrow}, \ b_{1\downarrow}, b_{2\downarrow}, b^{\dagger}_{1\downarrow}, b^{\dagger}_{2\downarrow}, \ d_{1\uparrow}, d_{2\uparrow}, d^{\dagger}_{1\uparrow}, d^{\dagger}_{2\uparrow}, \ d_{1\downarrow}, d_{2\downarrow}, d^{\dagger}_{1\downarrow}, d^{\dagger}_{2\downarrow}) \qquad (5.1)$$

Note that the space has four quartets of vectors.[34] This arrangement will be found to persist to the level of the dimension and fermion arrays of NEWQUeST, the octonion spectrum space of our universe. The subdivisions appear in the known fermion and internal symmetry fundamental representations as seen in Figs. 2b.1 and 2b.2.

5.1.2 Matrix Form of CASe Transformations

The form a CASe transformation can be symbolized by

---

[32] This space has space-time dimension 4 due to the compaction of two space-time dimensions.

[33] Some examples are the Kerr solution, space-times with gravity waves, and and spherically symmetric space-times.

[34] Note the form of eq. 5.1 appears to be complex-valued. But it is equivalent to a real-valued set.

$$x' = Tx \qquad (5.2)$$

where T is a 16×16 su(4, 4) transformation, and x and x' are 16-vectors in the fundamental representation of su(4, 4). In CASe the vectors are creation/annihilation operators that are transformed due to a GR transformation.

5.1.3 CASe Transformations for su(4, 4) over Sedenion Hypercomplex Numbers

CASe transformations take sixteen creation/annihilation operator terms (eq. 5.1) of eqs. 4.1 and 4.2, and generate 16 terms of the form of 4.26. We now extend CASe transformations to su(4, 4) over sedenion numbers. Thus each of the 16 terms in eqs. 4.1 and 4.2 becomes a sedenion number.

The use of sedenions, with their 16 parts, enables us to treat sets of 16 creation/annihilation operators as a unit as eqs. 5.1, 4.1, 4.2 and 4.26 suggest.

The vectors e_i are the 16 unit sedenions. They are numbered from 0 through 15. A sedenion S has the form

$$S = a_0 e_0 + a_1 e_1 + a_2 e_2 + \ldots + a_{15} e_{15} \qquad (5.3)$$
$$\equiv (a_0, a_1, \ldots, a_{15})$$

We now define the 16 sedenion terms as

$$\overset{i^{th}}{S_i = (0, 0, \ldots x_i, 0, 0, \ldots)} \qquad (5.4)$$

for i = 0, 1, 2, ..., 15 where x_i is the ith entry in eq. 5.1 above. We form a 16-vector

$$x = (S_0, S_1, \ldots, S_{15}) \qquad (5.5)$$

We define an su(4, 4) transformation over sedenions with

$$T = [a_{ij}] \qquad (5.6)$$

where

$$a_{ij} = s_{ij} e_i e_j = s_{ij} f_{ijk} e_k$$

where the s_{ij} are coefficients. The sedenion multiplication table appears in Fig. 5.1. Sedenions have an identity

$$e_i e_j = f_{ijk} e_k \qquad (5.7)$$

for integers f_{ijk} which are implicit in the table.

An su(4, 4) sedenion transformation has the form

$$x' = Tx \qquad (5.8)$$

and generates a sedenion 16-vector with each term of the form of eq. 4.26.

×	e_0	e_1	e_2	e_3	e_4	e_5	e_6	e_7	e_8	e_9	e_{10}	e_{11}	e_{12}	e_{13}	e_{14}	e_{15}
e_{15}	e_{15}	$-e_{14}$	e_{13}	e_{12}	$-e_{11}$	$-e_{10}$	e_9	$-e_8$	e_7	$-e_6$	e_5	e_4	$-e_3$	$-e_2$	e_1	e_0
e_{14}	e_{14}	e_{15}	$-e_{12}$	$-e_{13}$	$-e_{10}$	e_{11}	$-e_8$	$-e_9$	e_6	e_7	$-e_4$	$-e_5$	$-e_2$	e_3	$-e_0$	$-e_1$
e_{13}	e_{13}	e_{12}	$-e_{15}$	e_{14}	$-e_9$	$-e_8$	$-e_{11}$	e_{10}	e_5	e_4	e_7	e_6	$-e_1$	$-e_0$	$-e_3$	e_2
e_{12}	e_{12}	$-e_{13}$	$-e_{14}$	$-e_{15}$	$-e_8$	e_9	e_{10}	e_{11}	e_4	$-e_5$	$-e_6$	$-e_7$	$-e_0$	e_1	e_2	e_3
e_{11}	e_{11}	e_{10}	$-e_9$	$-e_8$	e_{15}	$-e_{14}$	e_{13}	$-e_{12}$	e_3	e_2	$-e_1$	$-e_0$	e_7	$-e_6$	e_5	$-e_4$
e_{10}	e_{10}	$-e_{11}$	$-e_8$	e_9	e_{14}	e_{15}	$-e_{12}$	$-e_{13}$	e_2	$-e_3$	$-e_0$	e_1	e_6	e_7	$-e_4$	$-e_5$
e_9	e_9	$-e_8$	e_{11}	$-e_{10}$	e_{13}	$-e_{12}$	$-e_{15}$	e_{14}	e_1	$-e_0$	e_3	$-e_2$	e_5	$-e_4$	$-e_7$	e_6
e_8	e_8	e_9	e_{10}	e_{11}	e_{12}	e_{13}	e_{14}	e_{15}	$-e_0$	$-e_1$	$-e_2$	$-e_3$	$-e_4$	$-e_5$	$-e_6$	$-e_7$
e_7	e_7	e_6	$-e_5$	$-e_4$	e_3	e_2	$-e_1$	$-e_0$	$-e_{15}$	$-e_{14}$	e_{13}	e_{12}	$-e_{11}$	$-e_{10}$	e_9	e_8
e_6	e_6	$-e_7$	$-e_4$	e_5	e_2	$-e_3$	$-e_0$	e_1	$-e_{14}$	e_{15}	e_{12}	$-e_{13}$	$-e_{10}$	e_{11}	e_8	$-e_9$
e_5	e_5	$-e_4$	e_7	$-e_6$	e_1	$-e_0$	e_3	$-e_2$	$-e_{13}$	e_{12}	$-e_{15}$	e_{14}	$-e_9$	e_8	$-e_{11}$	e_{10}
e_4	e_4	e_5	e_6	e_7	$-e_0$	$-e_1$	$-e_2$	$-e_3$	$-e_{12}$	$-e_{13}$	$-e_{14}$	$-e_{15}$	e_8	e_9	e_{10}	e_{11}
e_3	e_3	$-e_2$	e_1	$-e_0$	$-e_7$	e_6	e_5	e_4	$-e_{11}$	e_{10}	$-e_9$	e_8	e_{15}	$-e_{14}$	e_{13}	$-e_{12}$
e_2	e_2	e_3	$-e_0$	$-e_1$	$-e_6$	$-e_7$	e_4	e_5	$-e_{10}$	$-e_{11}$	e_8	e_9	e_{14}	e_{15}	$-e_{12}$	$-e_{13}$
e_1	e_1	$-e_0$	$-e_3$	e_2	$-e_5$	e_4	e_7	$-e_6$	$-e_9$	e_8	e_{11}	$-e_{10}$	e_{13}	$-e_{12}$	$-e_{15}$	e_{14}
e_0	e_0	e_1	e_2	e_3	e_4	e_5	e_6	e_7	e_8	e_9	e_{10}	e_{11}	e_{12}	e_{13}	e_{14}	e_{15}

Figure 5.1. The sedenion multiplication table. We have divided it into 4×4 parts reflecting the divisions of the fundamental fermions and the subgroups of Cayley-Dickson number n = 4 space (our universe).

The sedenion x′ is the result of a CASe transformation instigated by a GR transformation. It has 256 parts. We can display it as an array:

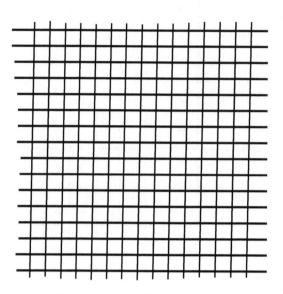

Figure 5.2. The sixteen sedenions of x′ expanded to form a 16 × 16 array of entries. Each row has the form of eq. 4.26.

In chapter 7 we will see that higher spaces also have a hypercomplex number aspect that emerges when GR in their space-time diimensions causes GiFT transformations of creation/annihilation operators.

6. Generation of Fermions and Internal Symmetries from a GR CASe Transformation

This chapter uses the form of GiFT transformations of creation/annihilation operators as a paradigm for transformations of fundamental fermions and internal symmetries to a non-static reference frame where the number of fermions is reduced from 256 to 16 and the internal symmetries are reduced to SU(3) $\otimes$U(1)$\otimes$SU(2)$\otimes$U(1) (or SU(4)$\otimes$SU(2)$\otimes$U(1)) forming a one generation Standard Model.

6.1 Fundamental Frame

A common feature of calculations in Physics is to choose a reference frame which facilitates computation. Common choices are the rest frame and the center of mass frame. In the preceding chapter we found that GR transformations to non-static reference frames induced creation/annihilation operator transformations. We further found that the transformations can be formulated using sedenion numbers. The result was transformations producing 16×16 arrays of numbers with a form like Fig. 5.2.

The appearance of sedenions and 16×16 arrays is analogous to the sedenions and dimension array of NEWQUeST in Octonion Cosmology. The 4×4 subblocks of 16 dimensions in NEWQUeST mirror the subblocks in Fig. 5.1 and the subblocking appearing in the creation/annihilation vector of eq. 5.1. Fig. 6.1 shows the similarity of the array structures in NEWQUeST and (2,2) Creation/Annihilation Space (CASe).

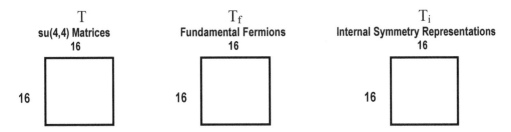

Figure 6.1 Comparison of arrays in CASe and NEWQUeST.

The analogous nature of these arrays raises the possibility that the fundamental fermion array and the Internal Symmetry array both are generated from a non-static reference frame via an su(4, 4) CASe-like transformation.

In chapter 5 we found

$$x' = Tx \qquad (5.8)$$

We now propose analogous transformations, T' and T'', that generate the fundamental fermion array and the Internal Symmetry array from a non-static coordinate system fermions and symmetries.

$$F_{ermions} = T'F_{fundamental} \qquad (6.1)$$
$$I_{QUeST} = T''I_{fundamental} \qquad (6.2)$$

Ideally $T' = T''$. In this approach $F_{fundamental}$ would be a sedenion vector composed of 16 sedenions of the form of eq. 5.4 where each sedenion has one primitive fermion. Thus it would represent 16 primitive fermions. Similarly $I_{fundamental}$ would be a sedenion vector composed of sedenions of the form of eq. 5.4 where each sedenion has one dimension. It would thus contain 16 real dimensions.

We call the reference frame for these reduced sets of fermions and dimensions the *Fundamental Frame*. It has a non-static coordinate system. (Almost all reference frames are non-static.)

The results of the transformations from the non-static Fundamental Frame are the NEWQUeST fermion and dimension arrays. Consequently the plethora of fermions and dimensions would be much reduced.

6.2 Contents of the Reduced Fermion and Internal Symmetry Arrays of the Fundamental Frame

The fermions and symmetries of the Fundamental Frame appear to mirror the Standard Model in part. On that basis we suggest the fermions appear in a 16-vector sedenion in analogy with eqs. 5.4 and 5.5. Each of the 16 sedenions contains one of the 16 fermions with each represented by an integer such as 1. This sedenion vector is mapped by a CASe transformation, eq. 6.1, to the 256 fermions of NEWQUeST and NEWUST.

Conceptually we can group the set of Fundamental Frame fermions into four fermion subblocks. Two of the subblocks contain Normal fermions. Two of the subblocks contain Dark fermions.

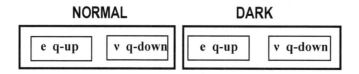

Figure 6.2. The 16 fermions of the Fundamental Frame grouped into Normal and Dark color SU(4) sets. Note each q-up and q-down are color triplets.

6.2.1 Four Types of Fermions

There are four types (species) of fermions: e-type, ν-type, up-quark type, and down-quark type. The species originate in a previous derivation of four fermion species (types) in the early 2000's. See Appendix D for the author's 2007 book *The Origin of the Standard Model* where it is derived in detail based on the four types of Complex Lorentz group boosts.

6.2.2 Transformation of Fundamental Frame Fermions to a "Normal" Frame

The proposed CASe transformation eq. 6.1 generates a 16-vector of sedenions in a "normal" static reference frame. Each sedenion component contains 16 fermions, which can be arranged to have the form of Fig. 6.3. We assume the arrangement of each of the output 16 sedenions, which contain numeric values, can map to the form of Fig. 6.2: Normal: e, q-up$_1$, q-up$_2$, q-up$_3$, v, q-down$_1$, q-down$_2$, q-down$_3$; Dark: e, q-up$_1$, q-up$_2$, q-up$_3$, v, q-down$_1$, q-down$_2$, q-down$_3$. Fig. 6.3 shows the 16 output sedenions displaying four layers with each containing four generations as in Fig. 6.4, and in NEWQUeST and NEWUST.

	NORMAL		DARK	
	e q-up	v q-down	e q-up	v q-down
	e q-up	v q-down	e q-up	v q-down
Layer 1	e q-up	v q-down	e q-up	v q-down
	e q-up	v q-down	e q-up	v q-down
	e q-up	v q-down	e q-up	v q-down
	e q-up	v q-down	e q-up	v q-down
Layer 2	e q-up	v q-down	e q-up	v q-down
	e q-up	v q-down	e q-up	v q-down
	e q-up	v q-down	e q-up	v q-down
	e q-up	v q-down	e q-up	v q-down
Layer 3	e q-up	v q-down	e q-up	v q-down
	e q-up	v q-down	e q-up	v q-down
	e q-up	v q-down	e q-up	v q-down
	e q-up	v q-down	e q-up	v q-down
Layer 4	e q-up	v q-down	e q-up	v q-down
	e q-up	v q-down	e q-up	v q-down

Figure 6.3. The 16-vector of sedenions with each containing 16 different fermions. Note the appearance of 4 layers and 4 generations of each fermion in each layer.

	Normal			Dark			Layer
	e q-up	**v q-down**		**e q-up**	**v q-down**		
	4	**4**		**4**	**4**		
4	□	□		□	□		**1**
4	□	□		□	□		**2**
4	□	□		□	□		**3**
4	□	□		□	□		**4**

Figure 6.4. Block form of NEWQUeST's and NEWUST's 256 fermions grouped by sixteens. Each block contains 4 generations of each fermion within the block.

6.2.3 Symmetry Groups within the Fundamental Frame – One Generation Standard Model

We provisionally assume that there are 16 symmetry group irreducible representation dimensions in a Fundamental Frame sedenion 16-vector in analogy with eqs. 5.4 and 5.5. Each of the 16 sedenions contains one of the 16 symmetry group dimensions in the form of an integer such as 1. This sedenion vector is mapped by a CASe transformation, eq. 6.1, to the 256 symmetry groups irreducible representations dimensions of NEWQUeST and NEWUST.

The 16 dimensions may be viewed as the dimensions of irreducible representations of U(8) broken to

$$SU(3)\otimes U(1) \quad SU(2)\otimes U(1)\otimes SL(2, \textbf{C})^{35}$$

In this case we can see a $SU(3)\otimes U(1)\otimes SU(2)\otimes U(1)\otimes SL(2, \textbf{C})$ set of symmetries arises giving a one generation Standard Model in 4 space-time dimensions. We have considered this type of Standard Model in the author's 2007 book *The Origin of the Standard Model*, which is reprinted in Appendix D.

The Fundamental Frame has the internal symmetry groups, the set of 8 normal fermions and 4-dimension space-time to form a Standard Model. The Standard Model, and its generalization to the Unified SuperStandard Theory, can be viewed as a

---

[35] We use SL(2, C) to represent $SO^+(1, 3)$.

transformation from the Fundamental Frame. The Fundamental Frame is non-static. There is an infinite set of possible Fundamental Frames since there is not a 1:1 correspondence between a General Relativistic transformation and a specific su(4, 4) transformation. A General Relativistic transformation corresponds to an infinite set of possible Fundamental Frames However the fermion and symmetry group contents of the Fundamental Frame is specific.

6.2.4 Transformation of Fundamental Frame Dimensions to a "Normal" Frame

The proposed CASe transformation eq. 6.1 generates a 16-vector of sedenions. Each sedenion component contains 16 irreducible representation dimensions with each in the form of a numeric value. A typical output sedenion has the form:

$$XXXXXXXXXXXXXXXX$$

where each x represents has some numeric value. The values may be the same or differ from entry to entry. We interpret these values as markers for irreducible symmetry group representation dimensions.

We arrange the 16 output sedenions in layers of Normal and Dark with the form of Fig. 6.5 with the form of each sedenion initially being 16 real U(4)⊗U(4) irreducible representation dimensions. Some sedenions are further subdivided into 8 real SU(4) irreducible representation dimensions,[36] and 4 real SU(2)⊗U(1) irreducible representation dimensions, and 4 real SL(2, **C**) space-time dimensions. See Figs. 6.5, 6.6. 6.7 and 6.8.

	NORMAL	**DARK**
	U(4)⊗U(4)	
	U(4)⊗U(4)	
Layer 1		U(4)⊗U(4)
		U(4)⊗U(4)
	U(4)⊗U(4)	
	U(4)⊗U(4)	
Layer 2		U(4)⊗U(4)
		U(4)⊗U(4)
	U(4)⊗U(4)	
	U(4)⊗U(4)	
Layer 3		U(4)⊗U(4)
		U(4)⊗U(4)
	U(4)⊗U(4)	
	U(4)⊗U(4)	
Layer 4		U(4)⊗U(4)
		U(4)⊗U(4)

Figure 6.5. The sixteen output dimension sedenions generated by a CASe transformation from the Fundamental Frame arranged in layers and generations.

---

[36] Or SU(3)⊗U(1) dimensions.

NORMAL		DARK	
SU(3)⊗U(1)	SU(2)⊗U(1)⊗SL(2, C)	SU(3)⊗U(1)	SU(2)⊗U(1)⊗SL(2, C)
Generation U(4)	Layer U(4)	Generation U(4)	Layer U(4)
SU(3)⊗U(1)	SU(2)⊗U(1)⊗SL(2, C)	SU(3)⊗U(1)	SU(2)⊗U(1)⊗SL(2, C)
Generation U(4)	Layer U(4)	Generation U(4)	Layer U(4)
SU(3)⊗U(1)	SU(2)⊗U(1)⊗SL(2, C)	SU(3)⊗U(1)	SU(2)⊗U(1)⊗SL(2, C)
Generation U(4)	Layer U(4)	Generation U(4)	Layer U(4)
SU(3)⊗U(1)	SU(2)⊗U(1)⊗SL(2, C)	SU(3)⊗U(1)	SU(2)⊗U(1)⊗SL(2, C)
Generation U(4)	Layer U(4)	Generation U(4)	Layer U(4)

Figure 6.6. The rearranged sixteen sedenions of Fig. 6.5 generated by a CASe transformation from the Fundamental Frame. This arrangement is the same as NEWQUeST and NEWUST.

6.3 A Basis for NEWQUeST and NEWUST in CASe Transformations from the Fundamental Frame

The set of internal symmetries of NEWQUeST and NEWUST emerges from Fig. 6.6. GiFT together with its General Relativistic aspect enables a reduction of the symmetries and fermions of NEWQUeST and NEWUST to 16 symmetry dimensions and 16 fermions.

The CASe transformations of GiFT, that provide a Fundamental Frame formulation of particles and interactions, *require* the PseudoQuantum formulation of GiFT. They implement a unification of Quantum theory and General Relativity since Quantum Theory is based on GiFT, which embodies Quantum Field Theory, which, in turn, embodies Quantum Mechanics.

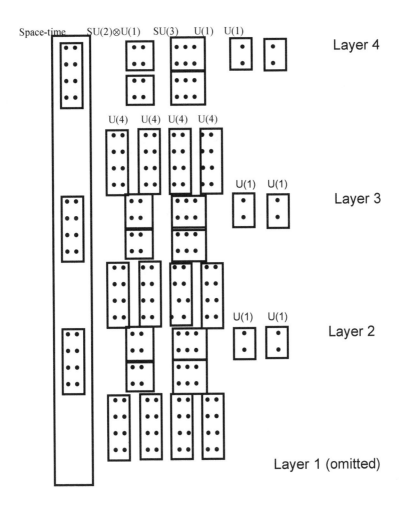

Figure 6.7. Three of the four layers of QUeST internal symmetry groups (and space-time) for Cayley-Dickson space 4. Layer 1 which has an identical form was omitted due to "page space" limitations. Note the left column of blocks combine to specify a 4 dimension octonion space-time. Note each layer has 64 dimensions.

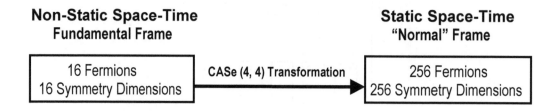

Figure 6.8. The CASe transformation from a Fundamental Frame to a "Normal" static space-time reference frame.

7. Higher Spaces CASe Transformations and Their Fundamental Frames

The higher Octonion Cosmology spaces where the Cayley-Dickson number is > 4 have Fundamental Frames that are analogous to the Fundamental Frame for n = 4 discussed in Chapter 6.

Figure 4.1 below displays the space-time dimension and the CASe group for these spaces. Since there is no name for hypercomplex numbers beyond sedenion, we will define *n-ion* to represent the hypercomplex Cayley-Dickson number n.

Each Octonion Cosmology n-ion space has a General Relativity. We call it n-ion GR. This chapter uses GiFT transformations of creation/annihilation operators as a paradigm for n-ion GR transformations of fermions and internal symmetries in an n-ion Fundamental Frame to a "normal" reference frame.

In an n-ion Fundamental Frame there are 2^n fermions and 2^n irreducible representation dimensions for internal symmetries. They are "boosted" to 2^{2n} fermions and 2^{2n} dimensions in a "normal" n-ion space reference frame by a CASe transformation. Correspondingly an n-ion GR transformation takes a Fundamental Frame to a "normal" reference frame.

The n-ion Fundamental Frame is non-static. There is an infinite set of possible n-ion Fundamental Frames since there is not a 1:1 correspondence between an n-ion General Relativistic transformation and a specific n-ion CASe transformation. An n-ion General Relativistic transformation corresponds to an infinite set of possible n-ion CASe transformations and thus to an infinite set of possible n-ion Fundamental Frames However the fermion and symmetry group contents of all the possible n-ion Fundamental Frames are the same.

CASe Groups of Eight of the Ten Octonion Universes

Octonion Space Number	Cayley-Dickson Number	Cayley Number	Dimension Array Total	Space-time-Dimension	Fermion Spinor Array Total	CASe Group $su(2^{r/2}, 2^{r/2})$
O_S	n	d_c	d_d	r	d_s	CASe
0	10	1024	1024×1024	18	512×512	su(512,512)
1	9	512	512×512	16	256×256	su(256,256)
2	8	256	256×256	14	128×128	su(128,128)
3	7	128	128×128	12	64×64	su(64,64)
4	6	64	64×64	10	32×32	su(32,32)
5	5	32	32×32	8	16×16	su(16,16)
6	4	16	16×16	$6 \rightarrow 4$	8×8	su(4,4)
7	3	8	8×8	4	4×4	su(4,4)

Figure 4.1. The CASe groups of the ten Octonion Cosmology spaces. Space n = 4 has 6 – 2 = 4 space-time dimensions due to 2 compacted dimensions.

We begin by noting the n-ion CASe group is $su(2^{n-1}, 2^{n-1})$, the total dimensions of the group representation is $2^n \times 2^n = 2^{2n}$, the space-time dimension is $r = 2n - 2$, and the dimension of the symmetry array of the Octonion Cosmology space is 2^{2n}.

In a manner similar to that of chapter 6 we see the CASe group dimensions equals the symmetry array dimensions 2^{2n}. See Fig. 7.1. On this basis this chapter uses n-ion GiFT CASe transformations of creation/annihilation operators as a paradigm for transformations between fermions and internal symmetries of a "normal" reference frame and an n-ion Fundamental Frame.

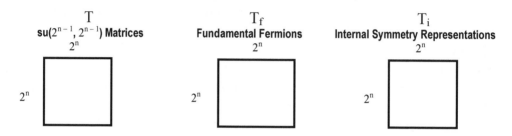

Figure 7.1 Comparison of arrays in CASe and NEWQUeST for the n-ion case.

$$F_{fermions} = T'F_{fundamental} \qquad (7.1)$$
$$I_{QUeST} = T''I_{fundamental} \qquad (7.2)$$

We assume $T' = T''$. Then $F_{fundamental}$ would be a n-ion vector composed of 2^n n-ions of the form of eq. 5.4 where each n-ion has one primitive fermion within it. The n-ion vector represents 2^n primitive fermions. Similarly $I_{fundamental}$ would be a n-ion vector composed of 2^n n-ion components of the form of eq. 5.4 where each n-ion component has one dimension. Thus the vector contains 2^n real-valued dimensions.

We call the reference frame containing these reduced sets of fermions and dimensions the *n-ion Fundamental Frame*. The results of the transformations from the non-static Fundamental Frame are the space n fermion and dimension arrays.

7.1 Contents of the Reduced Fermion and Internal Symmetry Arrays of the Fundamental Frame

We choose the fermions to appear in a 2^n-vector of n-ions in analogy with eqs. 5.4 and 5.5. Each of the 2^n n-ions contains one of the 2^n fermions with each represented by a number such as 1. This n-ion vector is mapped by a CASe transformation, eq. 7.1, to a set of n-ion vectors containing a total of 2^{2n} fermions in space n in a "normal" coordinate system. *Thus 2^n n-ion Fundamental Frame fermions become 2^{2n} fermions in the "normal" reference frame.*

Conceptually we can group the set of n-ion Fundamental Frame fermions into a set of 16 fermion subblocks with each containing the four types (species) of fermions: e-type, ν-type, up-quark type, and down-quark type. There are 2^{n-4} such subblocks.

7.1.1 Transformation of n-ion Fundamental Frame Fermions to a "Normal" Frame

A CASe transformation corresponding to eq. 7.1 generates a 2^n-vector of n-ions. Each n-ion component contains 2^n fermions expressed as numeric values. The arrangement of each of the 2^n output n-ions can be chosen to have the form of a multiple of 16 fermion subblocks: Normal: e, q-up$_1$, q-up$_2$, q-up$_3$, ν, q-down$_1$, q-down$_2$, q-down$_3$; Dark: e, q-up$_1$, q-up$_2$, q-up$_3$, ν, q-down$_1$, q-down$_2$, q-down$_3$. The 2^{2n} n-ion fermions then consist of 2^{2n-4} subblocks of a form similar to the form of the fermions in NEWQUeST and NEWUTMOST. See Figs. 6.2 – 6.4 for the NEWQUeST case.

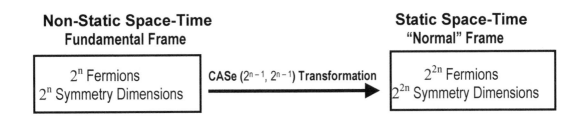

Figure 7.2. The CASe transformation from a non-static Fundamental Frame to a "Normal" static space-time reference frame.

7.1.2 Symmetry Groups within the n-ion Fundamental Frame

There are 2^n symmetry group irreducible representation dimensions in an n-ion Fundamental Frame n-ion-vector in analogy with eqs. 5.4 and 5.5. Each of the 2^n n-ions contains one of the 2^n symmetry group dimensions in the form of an integer such as 1. This n-ion vector is mapped by a CASe transformation, eq. 7.1, to 2^{2n} symmetry groups irreducible representations dimensions of space n.

7.1.3 Transformation of Fundamental Frame Dimensions to a "Normal" Frame

The CASe transformation eq. 7.1 generates a 2^n-vector of 2^n-n-ions from a Fundamental Frame 2^n-vector with one dimension in each n-ion. Each output n-ion component contains 2^n irreducible representation dimensions with each in the form of a numeric value. The 2^{2n} resulting dimensions are the symmetries' irreducible representations dimensions of space n. They represent the symmetries in Figs. 6.5 - 6.7 for the NEWQUeST case.

The n = 5 with su(16, 16) corresponds to NEWUTMOST, our proposed Megaverse space. Its 5-ion Fundamental Frame has 32 fermions and 32 symmetry group dimensions for

$$SU(3)\otimes U(1)\otimes SU(2)\otimes U(1)\otimes SL(2,\mathbf{C})\otimes SU(3)\otimes U(1)\otimes SU(2)\otimes U(1)\otimes SL(2,\mathbf{C})$$

presumably giving interactions to both the Normal and Dark fermion sectors.

The n = 8 CASe group case is interesting because its Fundamental Frame has 256 fermions and 256 symmetry dimensions. NEWQUeST has similar sets of fermions and dimensions. The n = 8 Fundamental Frame may contain an analogue to

NEWQUeST. This case raises the, perhaps remote, possibility that our universe has a non-static reference frame in the large.

The n = 10 CASe group case is interesting because its Fundamental Frame has 1024 fermions and 1024 symmetry dimensions. NEWUTMOST for the Megaverse has similar sets of fermions and dimensions. The n = 10 Fundamental Frame may contain an analogue to the NEWUTMOST Megaverse.

7.2 A Basis for NEWQUeST and NEWUST in CASe Transformations from Fundamental Frames

The set of internal symmetries of space n emerges. Again GiFT together with its General Relativistic aspect enables a reduction of the symmetries and fermions to 2^n symmetry dimensions and 2^n fermions in a non-static Fundamental Frame with a non-static reference frame.

The CASe transformations of GiFT, that provide an n-ion Fundamental Frame formulation of particles and interactions, again require the PseudoQuantum formulation of GiFT. They implement a unification of Quantum theory and General Relativity since Quantum Theory is based on GiFT, which embodies Quantum Field Theory, which, in turn, embodies Quantum Mechanics.

GiFT embodies General Relativity (and Special Relativity) in its PseudoQuantum formalism. This formalism supports a *canonical* formulation of higher derivative gravity with three distance scales that appears to give similar results to the MoND formulation of gravity. The PseudoQuantum formalism also supports higher derivative strong interactions with explicit quark confinement.[37] *NEWUST relates the three scale parameters of Gravitation to the coupling constant of the higher derivative, quark confining strong linear potential.*[38]

Gravitation joins the other gauge theories of interactions. It has a finite perturbation theory due to GiFT's Two-Tier Theory and PseudoQuantum Theory. Due to its long range nature (like QED) Gravitation has interesting solutions such as Black Holes just as Electrodynamics has interesting solutions that are used in Solid State Physics.

---

[37] Blaha (2018e) pages 460 – 465 and chapters 23 and 24.
[38] Chapters 22 – 24 in Blaha (2018e).

Appendix B. Some PseudoQuantum Papers

These papers appear in Blaha (2021j) as well as the journals.

S. Blaha, Phys. Rev. **D17**, 994 (1978).

S. Blaha, "The Local Definition of Asymptotic Particle States", IL Nuovo Cimento **49A**, 35 (1979).

S. Blaha, "New Framework for Gauge Field Theories", IL Nuovo Cimento **49A**, 113 (1979).

PseudoQuantum Theory of Color Confinement

S. Blaha, Phys. Rev. **D10**, 4268 (1974).

S. Blaha, Phys. Rev. D11, 2921 (1975).

Appendix C. Two-Tier and PseudoQuantum Formalisms

We are familiar with Quantum Theory, as it is usually developed, in our universe. In universes with higher space-time dimensions there is a need for a more expansive Quantum Field Theory. We presented this theory in earlier books on Two-Tier Quantum Theory[39] and PseudoQuantum Field Theory.[40] It resolved all divergences in ElectroWeak and Strong interaction theories. It also eliminated divergences in other types of Quantum Field Theories including theories with higher order derivatives and four fermion interactions.

In this book we focus on the higher space-time dimension spaces (universes) of Octonion Cosmology. In these universes we find Two-Tier Quantum Theory is *absolutely* necessary to avoid divergences in perturbation theory calculations. The universes of Octonion Cosmology have Feynman propagators and Perturbation Theory terms with integrations of the form:

$$\int d^n k \, f(k) \qquad (C.1)$$

where n = 4, 8, 10, 12, 14, 16, and 18 (and also a less interesting case n = 2.)

In eight space-time dimensions the second order single fermion loop vacuum polarization is sextic divergent: $\int d^8 k / k^2 \sim k^6$. Two-Tier Quantum Theory eliminates divergences in Perturbation Theory with exponentiated Gaussian quadratic terms in momenta of the form:

$$\exp(-ak^2) \qquad (C.2)$$

where a is a constant. Two-Tier Quantum Theory is required for Perturbation Theory calculations in higher dimensions.

PseudoQuantum Field Theory is also needed in Octonion Cosmology. It is needed for proper quantization in arbitrary coordinate systems that might be relevant in higher dimension Octonion Cosmology spaces. For example, consider coordinate systems for non-static space-times where no time-like Killing vector exists.[41] It also enables canonical higher order derivative theories for quark confinement and other purposes. And it dovetails with Two-Tier Coordinates to "dress" bare fermion and boson particles.

---

[39] See Blaha (2005a) *Quantum Theory of the Third Kind.*
[40] Blaha (2018e) and earlier books.
[41] B. DeWitt, Phys. Rep. **19**, 295 (1975) and references therein. S. Blaha, "New Framework for Gauge Field Theories", IL Nuovo Cimento **49A**, 113 (1979).

C.1 Reasons For Two-Tier Quantum Theory

Originally Two-Tier coordinates were developed by this author to remove infinities that appear in perturbation theory calculations. We showed that the quantum smeared coordinates of Two-Tier Quantum Field Theory succeeded in removing all ultra-violet infinities in perturbation theory including the fermion triangle infinities.

Remarkably the high precision, "low" energy[42] predictions of QED remained true in Two-Tier QED and thus remained consistent with experiment to a hitherto unsurpassed level of accuracy. "Low" energy predictions in other quantum field theories also remained unchanged. At high energies, Two-Tier perturbation theory results are finite and consequently all ultra-violet infinities, to any order in perturbation theory, in *any number of space-time dimensions* were eliminated.

In addition to removing perturbation theory infinities, Two-Tier coordinates enable us to define finite theories of Quantum Gravity and 'non-renormalizable' quantum field theories based on polynomial Lagrangians, to tame vacuum fluctuations, to eliminate infinities associated with the Big Bang, and possibly to generate the explosive growth of the universe in its role as a type of Dark Energy.[43]

C.2 Two-Tier Features in 4-Dimensional Space-Time

Two-Tier Quantum Field Theory[44] is based on a new method[45] in the Calculus of Variations that uses two 'layers' of fields to introduce quantum coordinates. We shall consider this technique for the specific case of a massless vector field $Y^\mu(y)$ where the index μ ranges from 0 through 3. It is.analogous to the electromagnetic field.

The X^μ coordinate system, where it appears, has a c-number real part and a q-number imaginary part. Thus particle fields which are normally defined on real four-dimensional real space-time will now be defined on a "slightly" complex four-dimensional space-time:

$$X^\mu(y) = y^\mu + i\, Y^\mu(y)/M_c^2 \qquad (C.3)$$

where M_c is an extremely large mass of the order of the Planck mass or larger.

The $Y^\mu(y)$ field is a function of the space y coordinates. The real part of the space-time dimensions will be taken to be the space of real-valued y coordinates.[46]

The imaginary part of space-time coordinates is the massless $Y^\mu(y)$ vector quantum field that is suppressed by the very large mass scale. The effects of Quantum Dimensions only become appreciable in quantum field theory at energies of the order of M_c. At these energies exponential Gaussian factors in each particle (and ghost) propagator are generated by the Quantum Dimensions and serve to make *all*

---

[42] Relative to a mass scale that was perhaps of the order of the Planck mass.

[43] See Blaha (2017b) and earlier books for details. This section is basically a summary of some features.

[44] See Blaha (2005a), and Blaha (2002), for discussions of this new method to eliminate infinities in quantum field theory calculations.

[45] See Blaha (2005a)..

[46] In a deeper theory the real part might also be a quantum field that undergoes a condensation to generate c-number coordinates. We will not consider this possibility in this book.

perturbation theory calculations ultra-violet finite – including calculations in Quantum Gravity. Later we will see that the Two-Tier formalism may be extended directly to the universes of Octonion Cosmology with similar results – finiteness in Perturbation Theory.

The Two-Tier formalism introduces a new form of interaction that does not have the form of the simple polynomial interactions that have hitherto dominated quantum field theories. This form of interaction takes place via the composition of quantum fields and can be called a *Dimensional Interaction* or an *Interdimensional Interaction* since it affects particle behavior through Quantum Dimensions.

The basic *ansatz* of the Two-Tier formalism is to replace every appearance of a coordinate x in a quantum field with the variable

$$x^\mu \to X^\mu = (y^0, \mathbf{y} + \mathbf{Y}(y^0, \mathbf{y})/M_c^2) \tag{C.4}$$

where $\mathbf{Y}(y^0, \mathbf{y})$ is the spatial part of a free massless vector field with features that are identical to the free QED field in the Radiation gauge.

Then one finds that the momentum space free field Feynman propagators G(k) of all particles acquires a Gaussian factor exp(h(k)):

$$G(k) \to G(k)\,\exp(h(k)) \tag{C.5}$$

so that all perturbation theory diagrams are finite. The result is finite perturbative results for all calculations to any order in perturbation theory. Blaha (2005a) shows that Two-Tier theories are finite, Poincare covariant, and unitary. (See Blaha (2005a) for a complete discussion.)

C.3 Two-Tier Quantum Coordinates Formalism

In this section we will introduce the basic Two-Tier formalism. Taking the Lagrangian described in Blaha (2005a):

$$\mathcal{L}(y) = \mathcal{L}_F\,(X^\mu(y))J + \mathcal{L}_C(X^\mu(y), \partial X^\mu(y)/\partial y^\nu, y) \tag{C.6}$$

where

$$X^\mu(y) = y^\mu + i\,Y^\mu(y)/M_c^2 \tag{C.7}$$

with M_c being a large mass scale, $Y_\mu(y)$ a vector quantum field, and where J is the absolute value of the Jacobian of the transformation from X to y coordinates:

$$J = |\partial(X)/\partial(y)|$$

The Lagrangian term $\mathcal{L}_C$ is

$$\mathcal{L}_C = +\tfrac{1}{4}\,M_c^4 F^{\mu\nu}F_{\mu\nu}$$

with

$$F_{\mu\nu} = \partial X_\mu/\partial y^\nu - \partial X_\nu/\partial y^\mu \tag{C.8}$$
$$\equiv i\,(\partial Y_\mu/\partial y^\nu - \partial Y_\nu/\partial y^\mu)/M_c^2$$

The Lagrangian term $\mathscr{L}_F(X^\mu(y))$ contains the terms for scalar, fermion and other gauge terms in general. The sign in $\mathscr{L}_C$ is not negative – contrary to the conventional electromagnetic Lagrangian. The reason for this difference is that the quantum field part of X^μ is imaginary. Thus $\mathscr{L}_C$ ends up having the correct sign after taking account of the factor of i in the field strength $F_{\mu\nu}$.

Defining

$$F_{Y\mu\nu} = (\partial Y_\mu/\partial y^\nu - \partial Y_\nu/\partial y^\mu)$$

we see the Lagrangian assumes the form of the conventional electromagnetic Lagrangian:

$$\mathscr{L}_C = -\tfrac{1}{4}\, F_Y^{\mu\nu} F_{Y\mu\nu}$$

The action of this theory has the form

$$I = \int d^4y\, \mathscr{L}(y)$$

Since $X^\mu(y)$ has an imaginary part there would appear to be an issue with the conservation of probability (unitarity). *We show unitarity is not a problem later in this appendix in section C.8.*

C.4 Y^μ Gauge

The gauge invariance of the Lagrangian allows us to choose a convenient gauge. The gauge invariance of the full Lagrangian:

$$\mathscr{L}_s = L_F(\phi(X), \partial\phi/\partial X^\mu)\, J + \mathscr{L}_C(X^\mu(y), \partial X^\mu(y)/\partial y^\nu)$$

is based on the standard gauge invariance of $\mathscr{L}_C$, and the gauge invariance of $J\mathscr{L}_F$ in the form of translational invariance

$$X^\mu(y) \to X^\mu(y) + \delta X^\mu(y)$$

for the special case of a translation of X with the form of a gauge transformation:

$$\delta X^\mu(y) = \partial\Lambda(y)/\partial y_\mu$$

In this case we find

$$\int d^4y\, \Lambda(y)\, \partial\, [\, J\, \partial/\partial X^\mu\, \mathscr{F}_{F\mu\nu}\,]/\partial y_\nu = 0 \qquad\qquad (C.9)$$

after a partial integration. Thus we have the differential conservation law:

$$\partial\, [\, J\, \partial\mathscr{F}_{F\mu\nu}/\partial X^\mu]/\partial y_\nu = 0$$

since $\Lambda(\mathbf{y})$ is arbitrary. This conservation law is trivially obeyed:

$$\partial \mathcal{T}_{F\mu\nu}/\partial X^{\mu} = 0 \tag{C.10}$$

Thus translational invariance in the $\mathcal{L}_F$ sector together with standard gauge invariance in the $\mathcal{L}_C$ sector automatically guarantees Y field gauge invariance of the total Lagrangian. We use the separate invariance of each term of

$$L = \int d^4 y \, [\mathcal{L}_F J + \mathcal{L}_C] = \int d^4 X \, \mathcal{L}_F + \int d^4 y \, \mathcal{L}_C = L_F + L_C$$

under a constant translation $X^{\mu} \rightarrow X^{\mu} + \delta X^{\mu}$ where δX^{μ} is constant. Then we consider a position dependent translation/gauge transformation, which taken together with the above equation, establishes the invariance under the position dependent translation/gauge transformation.

An alternate approach that leads to the same result is to start with the particle part of the Lagrangian $\mathcal{L}_f$ rewritten to be invariant under general coordinate transformations, as it must, when we generalize to include General Relativity. Since position dependent translations are a form of general coordinate transformation the full theory must be invariant under position dependent translations due to invariance under general coordinate transformations.

Having established invariance under gauge transformations we now choose to use the most convenient gauge – the radiation gauge[47]:

$$\partial Y^i/\partial y^i = 0 \tag{C.11}$$

where i = 1, 2, 3. In the absence of external sources, we set

$$Y^0 = 0$$

since Y^0 does not have a canonically conjugate momentum. A conventional treatment leads to the equal time commutation relations:

$$[Y^{\mu}(\mathbf{y}, y^0), Y^{\nu}(\mathbf{y}', y^0)] = [\pi^{\mu}(\mathbf{y}, y^0), \pi^{\nu}(\mathbf{y}', y^0)] = 0 \tag{C.12}$$
$$[\pi_j(\mathbf{y}, y^0), Y_k(\mathbf{y}', y^0)] = -i \, \delta^{tr}_{jk}(\mathbf{y} - \mathbf{y}')$$

where (Note the locations of the j indexes above introduce a minus sign.)

$$\pi^k = \partial \mathcal{L}_C/\partial Y_k'$$

---

[47] It is also possible to quantize using an indefinite metric that preserves manifest Lorentz covariance as was done by Gupta and Bleuler for the electromagnetic field. We will use the Gupta-Bleuler approach later to establish covariance under special relativity later. Now we opt for manifest positivity and use the radiation gauge.

$$\pi^0 = 0$$

$$\delta^{tr}_{jk}(\mathbf{y} - \mathbf{y}') = \int d^3k \, e^{i \, \mathbf{k} \cdot (\mathbf{y} - \mathbf{y}')}(\delta_{jk} - k_j k_k / \mathbf{k}^2)/(2\pi)^3 \qquad (C.13)$$

$$Y_k' = \partial Y_k / \partial y^0$$

The Radiation gauge reveals the two degrees of freedom that are present in the vector potential. The Fourier expansion of the vector potential is:

$$Y^i(y) = \int d^3k \, N_0(k) \sum_{\lambda=1}^{2} \varepsilon^i(k, \lambda)[a(k,\lambda) \, e^{-ik \cdot y} + a^\dagger(k,\lambda) \, e^{ik \cdot y}] \qquad (C.14)$$

where

$$N_0(k) = [(2\pi)^3 2\omega_k]^{-\frac{1}{2}}$$

and (since m = 0)

$$\omega_k = (\mathbf{k}^2)^{\frac{1}{2}} = k^0$$

with $\vec{\varepsilon}(k, \lambda)$ being the polarization unit vectors for $\lambda = 1,2$ and $k^\mu k_\mu = 0$.

The further development of Two-Tier theory is described in Part 3 of Blaha (2005a).

C.5 Two-Tier Uncertainty Principle

The Uncertainty Relation for Quantum Mechanics is based on the coordinate-momentum commutator. Similarly, in defining Quantum Coordinates we have established a commutation relation based on Y^μ. Its conjugate momentum is

$$P^\mu(y) = i\pi_Y{}^\mu(y)/M_c^2 \qquad (C.15)$$

where

$$\pi_Y{}^\mu(y) = - \, dY^\mu/(y)dy^0$$

In the Radiation gauge (eq. C.11) we see

$$\pi_Y{}^0(y) = 0 \qquad (C.16)$$

and

$$[X^0, P^0] = [y^0, p^0] = 0 \qquad (C.17)$$

The non-zero equal time commutation relation expressing a quantum Uncertainty Relation is

$$[P^i(\mathbf{y}, y^0) , Y^k(\mathbf{y}', y^0)] = -[\pi_Y{}^i(y), Y^k(y),]/M_c^4 \qquad (C.18)$$

$$= i\delta^{trik}(\mathbf{y} - \mathbf{y'})/M_c^4$$

using eq. C.12.

At low energy the impact of the Two-Tier Uncertainty Relation is diminished (more or less eliminated) by the factor M_c^4. Then Two-Tier Quantum Field Theory becomes ordinary Quantum Field Theory with the same results in Perturbation Theory. This limit is described in detail in Blaha (2005a), which appears in in Appendix B.

In the "low" energy limit the conventional Heisenberg Uncertainty Condition becomes evident as shown by Heitler (1954) and others. *Conventional Quantum Mechanics is a result of Second Quantization.*

C.6 Quantum Mechanics vs. Quantum Field Theory vs. Two-Tier Quantum Theory

Historically, Quantum Mechanics predated Quantum Field Theory, which, in turn, predates the Two-Tier Quantum Theory developed by the author in the early 2000's.

Logically, Two-Tier Quantum Theory, which embodies an Uncertainty Relation at ultrahigh energies, is the fundamental Quantum Theory. It is the predecessor of Quantum Field Theory, which only embodies the Heisenberg Uncertainty Relation. Quantum Field Theory is flawed by high energy divergences just as atomic physics, before the Bohr atom and Quantum Mechanics, was flawed by infinities in hydrogen atom models.

Quantum Field Theory (a "low" energy theory) implies Quantum Mechanics[48] as shown in Heitler[49] as well as elsewhere.

Quantum Mechanics is often treated as a complete, self-contained theory. On occasion paradoxes and ambiguities are found in Quantum Mechanics studies. They may be resolved within Quantum Mechanics. Any, that are not so resolved, should be considered within the framework of Quantum Field Theory, which is the ultimate forum for all quantum phenomena at "low" energy.

Note the trend of quantum theories, from the earliest at "low" energy theory, Quantum Mechanics, to the highest energy quantum theory: Two-Tier Quantum Theory,

Two-Tier Quantum Field Theory is a welcome extension of Quantum Theory to the deepest levels of Quantum Theory.

C.7 Two-Tier Perturbation Theory

The form of Two-Tier Perturbation Theory is similar to the Perturbation Theory of conventional Quantum Field Theory. It is described in *Quantum Theory of the Third Kind.* The reduction development for the U-matrix and the S-matrix are presented there.

---

[48] Quantum Mechanics is often treated as an independent theory. This practice may be valid mathematically but it is not valid physically. For example, many phenomena in atomic theory require explanation in terms of Quantum Field Theory.

[49] Heitler (1954).

C.8 Two-Tier Unitarity

The unitarity of Two-Tier Quantum Field Theory can be viewed in the cases of "low" energy phenomena and "high" energy phenomena with the scale mass of M_c separating the cases.

In the case of "low" energy phenomena the Two-Tier S-matrix gives results identical to conventional S-matrix results. Thus there is no conflict with unitarity for Two-Tier S-matrix results at "low" energy.

At "high" energy there are two potential issues: asymptotic states containing Y quanta; and conservation of probability. The first issue is resolved in the chapter 6 discussion in Appendix B of Blaha (2021i).

The second issue leads to a requirement to renormalize S-marix probability amplitudes such that their absolute values squared sum to unity – the unitarity condition. If we let S-matrix elements have the form S_{fi} where i represents an initial state and f represents one of its final states, then

$$\sum_n S_{nf}^* S_{ni} = \delta_{fi} \qquad (C.19)$$

expresses the conventional unitarity condition. In Two-Tier Quantum Field Theory the unitarity condition for energies much less than M_c is the same as eq. C.19 since perturbation theory results are the same as conventional Quantum Field Theory. The sum over intermediate states n is restricted to states whose total energy is less than M_c by energy conservation.

For initial states i with energy E_i of the order of or greater than M_c, the intermediate state total energy and the final state total energy are also E_i by energy conservation. In this case the sum in eq. C.19 *may* be

$$\sum_n S_{nf}^* S_{ni} = g_i \delta_{fi} \qquad (C.20)$$

where g_i is a constant term dependent on the state i. In this circumstance unitarity may be recovered by redefining S_{ni} and S_{nf} as

$$S_{ni}' = S_{ni}/\sqrt{g_i} \qquad (C.21)$$
$$S_{nf}' = S_{nf}/\sqrt{g_f} \qquad (C.22)$$

Then the *normalized* Two-Tier unitarity condition can be expressed as

$$\sum_n S_{nf}'^* S_{ni}' = \delta_{fi} \qquad (C.23)$$

with the understanding that $g_i = 1$ for initial states with energies much less than M_c. Thus Two-Tier Quantum Field Theory satisfies an enhanced unitarity condition.

C.9 Two-Tier Quantum Space Theory

Quantum Space Theory is the theory of particles containing spaces internally. It was recently developed by the author.[50] This theory can directly put in the form of Two-Tier Quantum Space Theory by following the procedure described in section C.2. The structure of the Octonion Spaces Spectrum is not changed by this formulation.

C.10 CQ Mechanics – A Union of Classical and Quantum Mechanics

The author developed a larger theory, CQ Mechanics, containing both Quantum Mechanics and Classical Mechanics that is described in Blaha (2016f). In this theory a phenomenon can be described as classical in one limit and as quantum in another limit. The description is determined by an angle that specifies one, or the other, limiting case. Intermediate cases combining both quantum and classical are also present. The theory appears here in Appendix C.

This theory can bridge the classical and quantum regimes. We discuss some applications in Blaha (20016f): a generalized Feynman path integral formalism, a generalized Schrödinger equation, a generalized Boltzmann equation, the Fokker-Planck equation, a generalized approach to quantum and classical chaos, and to quantum entanglement as well as semi-quantum entanglement. Our formalism applies to Quantum Mechanics as well as the path integrals, the Fokker-Planck equation and the Boltzmann equation.

This "Mechanics" theory has an analogue at the field theoretic level, PseudoQuantum Field Theory that is briefly described next.

C.11 PseudoQuantum Field Theory

PseudoQuantum Field Theory[51] was developed by the author in the 1970's and presented in a series of articles including the papers below.

S. Blaha, "The Local Definition of Asymptotic Particle States", IL Nuovo Cimento **49A**, 35 (1979). It describes the PseudoQuantization of boson and fermion field theories for use in the quantization of fields in universes and the Megaverse.

S. Blaha, "New Framework for Gauge Field Theories", IL Nuovo Cimento **49A**, 113 (1979). It describes the PseudoQuantization of gauge field theories for the purposes of defining higher derivative field theories and for use in the quantization of fields in universes and the Megaverse.

PseudoQuantum Field Theory (and its Quantum Mechanics analogue CQ Mechanics[52]) originated in the need to second quantize in unusual coordinate systems, and in curved space-time coordinate systems. It is also very relevant for the canonical formulation of higher derivative field theories for quark confinement and other applications.

---

[50] Blaha (2021f) and (2021g).
[51] Appendix C describes PseudoQuantum Theory features in some detail.
[52] See Blaha (2016f) for details, which contains Blaha (2016f). CQ Mechanics encompasses both classical mechanics and quantum mechanics, and provides a method of rotating between them. It has applications to transitions between Quantum/Semi-Classical Entanglement, and Quantum/Classical Path Integrals, and Quantum/Classical Chaos.

PseudoQuantum Field Theory is formulated by duplicating all fields, both fermion and boson fields, in a "normal" Lagrangian theory. Scalar field theory provides a simple example that illustrates the PseudoQuantum Field Theory procedure.

We duplicate a scalar field generating two scalar fields: $\varphi_1(x)$ and $\varphi_2(x)$. We choose $\varphi_1(x)$ to have a zero equal time commutator with $d\varphi_1(x)/dx^0$ and $\varphi_2(x)$ to have a conventional equal time commutator with $d\varphi_2(x)/dx^0$. Conceptually $\varphi_1(x)$ is a "classical" field and $\varphi_2(x)$ is a quantum field. A Lagrangian that implements these choices of commutation relations is:

$$\mathcal{L} = \partial^\mu \varphi_1(x)\partial_\mu\varphi_2(x) - \tfrac{1}{2}\,\partial^\mu \varphi_1(x)\partial_\mu\varphi_1(x) - m_2{}^2\,\varphi_1(x)\varphi_2(x) + \tfrac{1}{2}\,m_1{}^2\,\varphi_1(x)^2 \qquad (C.24)$$

$$(\Box + m_2{}^2)\varphi_1(x) = 0 \qquad (C.25)$$

$$(\Box + m_2{}^2)\varphi_2(x) - (\Box + m_1{}^2)\varphi_1(x) \quad = 0 \qquad (C.26)$$

The canonical momenta are

$$\pi_1 = d\varphi_2(x)/dt - d\varphi_1(x)/dt \qquad (C.27)$$
$$\pi_2 = d\varphi_1(x)/dt \qquad (C.28)$$

and the equal time commutation relations are

$$[\varphi_i(x), \pi_j(y)] = i\delta_{ij}\delta(\mathbf{x} - \mathbf{y}) \qquad (C.29)$$
$$[\varphi_i(x), \varphi_j(y)] = [\pi_i(x), \pi_j(y)] = 0 \qquad (C.30)$$

implying

$$[\varphi_1(x), d\varphi_1(y)/dt] = 0 \qquad (C.31)$$
$$[\varphi_2(x), d\varphi_2(y)/dt] = i\delta(\mathbf{x} - \mathbf{y}) \qquad (C.32)$$
$$[\varphi_1(x), d\varphi_2(y)/dt] = i\delta(\mathbf{x} - \mathbf{y}) \qquad (C.33)$$

C.11.1 Color Confinement

Appendix F contains a paper on a non-Abelian gauge field theory that gives explicit color confinement using a PseudoQuantum formulation that leads to a confining r-potential.

If we set $m_2 = 0$ in the above example, we see a fourth order field equation results

$$\Box^2\varphi_2(x) = 0 \qquad (C.34)$$

Color confinement results in the non-Abelian gauge field case in a similar manner.

C.11.2 PseudoQuantum Quantization for Non-Static Coordinate Systems

The PseudoQuantum formalism that was developed by the author for "normal" and non-static coordinate systems in the 1970s.

C.11.3 Advantages of PseudoQuantum Quantization

In this section we point out some of its advantages in a variety of field theory contexts that are relevant for Octonion Cosmology and Quantum Field Theory in general.

Some advantages of PseudoQuantum Field Theory are:

1. Quantization in any coordinate system in flat or curved space-times with an invariant definition of asymptotic particle states. *This is especially important for the higher dimension spaces of Octonion Cosmology.* An n particle asymptotic state in one coordinate system is a unitarily equivalent n particle asymptotic state in any other coordinate system. Therefore particle number is invariant under change of coordinate system. This is important for the Unified SuperStandard Theory in curved space-times. It is also important for quantization in higher dimensional spaces such as those of Octonion Cosmology. The method was developed in the late 1970's by the author to provide a quantization procedure which supports a unique particle interpretation of states in arbitrary non-static space-times where no global time-like coordinate (Killing vector) exists. PseudoQuantum Field Theory which we developed in a series of books[53] also can be formulated in the Octonion Spectrum of spaces. For example, we can use it to implement the Higgs Mechanism to generate particle masses and symmetry breaking.

2. PseudoQuantum Field Theory enables one to define Higgs particle dynamics in such a way that a non-zero vacuum expectation value cleanly separates from the quantum field part of the Higgs fields. This technique can be used in symmetry breaking mechanisms, mass generation, and possible generation of coupling constants as vacuum expectation values.

3. It supports the canonical definition of higher derivative field theories through the use of the Ostrogradski bootstrap. We have used it to construct a fourth order theory of the Strong interaction that has color confinement and a linear r potential. The potential part of this theory was used by the Cornell group to calculate the Charmonium spectrum. (See Blaha (2017b) for details.)

C.12 Combination of Two-Tier and PseudoQuantum Formalisms

These formalisms can be directly combined with no issues.

C.13 A Model Illustrating Scalar Field Quantization Using X^μ

We begin by considering the case of a scalar quantum field theory. We assume a real underlying y subspace. Since X^μ is a set of coordinates, we choose to define a scalar field ϕ as a function of X^μ, which, in turn, is a function of the y^ν coordinates. We will

---

[53] See Blaha (2017b) for the discussion of the PseudoQuantum field theory formalism for Higgs particles in our Extended Standard Model. See chapter 20 of Blaha (2017b), and earlier books, for a more detailed view than that presented here.

provisionally second quantize ϕ treating X^μ as c-number coordinates using a conventional approach.[54]

We assume a Lagrangian, with the momentum conjugate to ϕ:

$$\pi_\phi = \partial L_F / \partial \phi' \equiv \partial L_F / \partial(\partial \phi / \partial X^0) \tag{C.35}$$

Following the canonical quantization procedure, π and ϕ become Hermitian operators with equal time ($X^0 = X^{0'}$) commutation rules:

$$[\phi(X), \phi(X')] = [\pi_\phi(X), \pi_\phi(X')] = 0 \tag{C.36}$$
$$[\pi_\phi(X), \phi(X')] = -i\,\delta^3(\mathbf{X} - \mathbf{X}')$$

The standard Fourier expansion of the solution to the Klein-Gordon equation is:

$$\phi(X) = \int d^3p \, N_m(p) \, [a(p) \, e^{-ip \cdot X} + a^\dagger(p) \, e^{ip \cdot X}] \tag{C.36a}$$
$$= \int d^{3k}p \, N_m(p) \, [a(p) \, \exp(-ip \cdot (y + Y/M_c^2)) +$$
$$+ a^\dagger(p) \, \exp(ip \cdot (y + Y/M_c^2))] \tag{C.36b}$$

where

$$N_m(p) = [(2\pi)^3 2\omega_p]^{-\frac{1}{2}}$$

and

$$\omega_p = (\mathbf{p}^2 + m^2)^{\frac{1}{2}}$$

The commutation relations of the Fourier coefficient operators are:

$$[a(p), a^\dagger(p')] = \delta^3(\mathbf{p} - \mathbf{p}')$$
$$[a^\dagger(p), a^\dagger(p')] = [a(p), a(p')] = 0$$

The reader will recognize the quantization procedure is formally identical to the standard canonical quantization procedure of a free scalar quantum field.

In the case of spin ½, spin 1 and spin 2 fields the standard quantization procedure *in terms of the X coordinate system* can also be followed in a way similar to the procedure in standard texts.

C.14 Scalar Feynman Propagators

The momentum space free field Feynman propagators G…(k) of all particles and ghosts in all Two-Tier Quantum Field Theories acquires a Gaussian factor $\exp(h(k))$:

---

[54] Some texts are: Bogoliubov, N. N., Shirkov, D. V., *Introduction to the Theory of Quantized Fields* (Wiley-Interscience Publishers Inc., New York, 1959); Bjorken, J. D., Drell, S. D., *Relativistic Quantum Fields* (McGraw-Hill, New York, 1965); Huang, K., *Quarks, Leptons & Gauge Fields Second Edition* (World Scientific, River Edge, NJ, 1992); Kaku, M., *Quantum Field Theory* (Oxford University Press, New York, 1993); Weinberg, S., *The Quantum Theory of Fields* (Cambridge University Press, New York, 1995).

$$G\ldots(k) \rightarrow G\ldots(k) \exp(h(k))$$

so that all perturbation theory diagrams are finite. The result is a finite perturbative result in all calculations to any order in perturbation theory. Blaha (2005a) shows that Two-Tier theories are finite, Poincare covariant, and unitary.

REFERENCES

Akhiezer, N. I., Frink, A. H. (tr), 1962, *The Calculus of Variations* (Blaisdell Publishing, New York, 1962).

Bjorken, J. D., Drell, S. D., 1964, *Relativistic Quantum Mechanics* (McGraw-Hill, New York, 1965).

Bjorken, J. D., Drell, S. D., 1965, *Relativistic Quantum Fields* (McGraw-Hill, New York, 1965).

Blaha, S., 1995, *C++ for Professional Programming* (International Thomson Publishing, Boston, 1995).

_____, 1998, *Cosmos and Consciousness* (Pingree-Hill Publishing, Auburn, NH, 1998 and 2002).

_____, 2002, *A Finite Unified Quantum Field Theory of the Elementary Particle Standard Model and Quantum Gravity Based on New Quantum Dimensions™ & a New Paradigm in the Calculus of Variations* (Pingree-Hill Publishing, Auburn, NH, 2002).

_____, 2004, *Quantum Big Bang Cosmology: Complex Space-time General Relativity, Quantum Coordinates,™ Dodecahedral Universe, Inflation, and New Spin 0, ½, 1 & 2 Tachyons & Imagyons* (Pingree-Hill Publishing, Auburn, NH, 2004).

_____, 2005a, *Quantum Theory of the Third Kind: A New Type of Divergence-free Quantum Field Theory Supporting a Unified Standard Model of Elementary Particles and Quantum Gravity based on a New Method in the Calculus of Variations* (Pingree-Hill Publishing, Auburn, NH, 2005).

_____, 2005b, *The Metatheory of Physics Theories, and the Theory of Everything as a Quantum Computer Language* (Pingree-Hill Publishing, Auburn, NH, 2005).

_____, 2005c, *The Equivalence of Elementary Particle Theories and Computer Languages: Quantum Computers, Turing Machines, Standard Model, Superstring Theory, and a Proof that Gödel's Theorem Implies Nature Must Be Quantum* (Pingree-Hill Publishing, Auburn, NH, 2005).

_____, 2006a, *The Foundation of the Forces of Nature* (Pingree-Hill Publishing, Auburn, NH, 2006).

_____, 2006b, *A Derivation of ElectroWeak Theory based on an Extension of Special Relativity; Black Hole Tachyons; & Tachyons of Any Spin.* (Pingree-Hill Publishing, Auburn, NH, 2006).

_____, 2007a, *Physics Beyond the Light Barrier: The Source of Parity Violation, Tachyons, and A Derivation of Standard Model Features* (Pingree-Hill Publishing, Auburn, NH, 2007).

_____, 2007b, *The Origin of the Standard Model: The Genesis of Four Quark and Lepton Species, Parity Violation, the ElectroWeak Sector, Color SU(3), Three Visible Generations of Fermions, and One Generation of Dark Matter with Dark Energy* (Pingree-Hill Publishing, Auburn, NH, 2007).

_____, 2008a, *A Direct Derivation of the Form of the Standard Model From GL(16)* (Pingree-Hill Publishing, Auburn, NH, 2008).

_____, 2008b, *A Complete Derivation of the Form of the Standard Model With a New Method to Generate Particle Masses Second Edition* (Pingree-Hill Publishing, Auburn, NH, 2008)

_____, 2009, *The Algebra of Thought & Reality: The Mathematical Basis for Plato's Theory of Ideas, and Reality Extended to Include A Priori Observers and Space-Time Second Edition* (Pingree-Hill Publishing, Auburn, NH, 2009).

_____, 2010a, *Operator Metaphysics: A New Metaphysics Based on a New Operator Logic and a New Quantum Operator Logic that Lead to a Mathematical Basis for Plato's Theory of Ideas and Reality* (Pingree-Hill Publishing, Auburn, NH, 2010).

_____, 2010b, *The Standard Model's Form Derived from Operator Logic, Superluminal Transformations and GL(16)* (Pingree-Hill Publishing, Auburn, NH, 2010).

_____, 2010c, *SuperCivilizations: Civilizations as Superorganisms* (McMann-Fisher Publishing, Auburn, NH, 2010).

_____, 2011a, *21*[st] *Century Natural Philosophy Of Ultimate Physical Reality* (McMann-Fisher Publishing, Auburn, NH, 2011).

_____, 2011b, *All the Universe! Faster Than Light Tachyon Quark Starships & Particle Accelerators with the LHC as a Prototype Starship Drive Scientific Edition* (Pingree-Hill Publishing, Auburn, NH, 2011).

_____, 2011c, *From Asynchronous Logic to The Standard Model to Superflight to the Stars* (Blaha Research, Auburn, NH, 2011).

_____, 2012a, *From Asynchronous Logic to The Standard Model to Superflight to the Stars volume 2: Superluminal CP and CPT, U(4) Complex General Relativity and The Standard Model, Complex Vierbein General Relativity, Kinetic Theory, Thermodynamics* (Blaha Research, Auburn, NH, 2012).

_____, 2012b, *Standard Model Symmetries, And Four And Sixteen Dimension Complex Relativity; The Origin Of Higgs Mass Terms* (Blaha Reasearch, Auburn, NH, 2012).

_____, 2013a, *Multi-Stage Space Guns, Micro-Pulse Nuclear Rockets, and Faster-Than-Light Quark-Gluon Ion Drive Starships* (Blaha Research, Auburn, NH, 2013).

_____, 2013b, *The Bridge to Dark Matter; A New Sister Universe; Dark Energy; Inflatons; Quantum Big Bang; Superluminal Physics; An Extended Standard Model Based on Geometry* (Blaha Reasearch, Auburn, NH, 2013).

_____, 2014a, *Universes and Megaverses: From a New Standard Model to a Physical Megaverse; The Big Bang; Our Sister Universe's Wormhole; Origin of the Cosmological Constant, Spatial Asymmetry of the Universe, and its Web of Galaxies; A Baryonic Field between Universes and Particles; Megaverse Extended Wheeler-DeWitt Equation* (Blaha Reasearch, Auburn, NH, 2014).

_____, 2014b, *All the Megaverse! Starships Exploring the Endless Universes of the Cosmos Using the Baryonic Force* (Blaha Research, Auburn, NH, 2014).

_____, 2014c, *All the Megaverse! II Between Megaverse Universes: Quantum Entanglement Explained by the Megaverse Coherent Baryonic Radiation Devices – PHASERs Neutron Star Megaverse Slingshot Dynamics Spiritual and UFO Events, and the Megaverse Microscopic Entry into the Megaverse* (Blaha Research, Auburn, NH, 2014).

_____, 2015a, *PHYSICS IS LOGIC PAINTED ON THE VOID: Origin of Bare Masses and The Standard Model in Logic, U(4) Origin of the Generations, Normal and Dark Baryonic Forces, Dark Matter, Dark Energy, The Big Bang, Complex General Relativity, A Megaverse of Universe Particles* (Blaha Research, Auburn, NH, 2015).

_____, 2015b, *PHYSICS IS LOGIC Part II: The Theory of Everything, The Megaverse Theory of Everything, U(4)$\otimes$U(4) Grand Unified Theory (GUT), Inertial Mass = Gravitational Mass, Unified Extended Standard Model and a New Complex General Relativity with Higgs Particles, Generation Group Higgs Particles* (Blaha Research, Auburn, NH, 2015).

_____, 2015c, *The Origin of Higgs ("God") Particles and the Higgs Mechanism: Physics is Logic III, Beyond Higgs – A Revamped Theory With a Local Arrow of Time, The Theory of Everything Enhanced, Why Inertial Frames are Special, Universes of the Mind* (Blaha Research, Auburn, NH, 2015).

_____, 2015d, *The Origin of the Eight Coupling Constants of The Theory of Everything: U(8) Grand Unified Theory of Everything (GUTE), S^8 Coupling Constant Symmetry, Space-Time Dependent Coupling Constants, Big Bang Vacuum Coupling Constants, Physics is Logic IV* (Blaha Research, Auburn, NH, 2015).

_____, 2016a, *New Types of Dark Matter, Big Bang Equipartition, and A New U(4) Symmetry in the Theory of Everything: Equipartition Principle for Fermions, Matter is 83.33% Dark, Penetrating the Veil of the Big Bang, Explicit QFT Quark Confinement and Charmonium, Physics is Logic V* (Blaha Research, Auburn, NH, 2016).

_____, 2016b, *The Periodic Table of the 192 Quarks and Leptons in The Theory of Everything: The U(4) Layer Group, Physics is Logic VI* (Blaha Research, Auburn, NH, 2016).

_____, 2016c, *New Boson Quantum Field Theory, Dark Matter Dynamics, Dark Matter Fermion Layer Mixing, Genesis of Higgs Particles, New Layer Higgs Masses, Higgs Coupling Constants, Non-Abelian Higgs Gauge Fields, Physics is Logic VII* (Blaha Research, Auburn, NH, 2016).

_____, 2016d, *Unification of the Strong Interactions and Gravitation: Quark Confinement Linked to Modified Short-Distance Gravity; Physics is Logic VIII* (Blaha Research, Auburn, NH, 2016).

_____, 2016e, *MoND: Unification of the Strong Interactions and Gravitation II, Quark Confinement Linked to Large-Scale Gravity, Physics is Logic IX* (Blaha Research, Auburn, NH, 2016).

_____, 2016f, *CQ Mechanics: A Unification of Quantum & Classical Mechanics, Quantum/Semi-Classical Entanglement, Quantum/Classical Path Integrals, Quantum/Classical Chaos* (Blaha Research, Auburn, NH, 2016).

_____, 2016g, *GEMS: Unified Gravity, ElectroMagnetic and Strong Interactions: Manifest Quark Confinement, A Solution for the Proton Spin Puzzle, Modified Gravity on the Galactic Scale* (Pingree Hill Publishing, Auburn, NH, 2016).

_____, 2016h, *Unification of the Seven Boson Interactions based on the Riemann-Christoffel Curvature Tensor* (Pingree Hill Publishing, Auburn, NH, 2016).

_____, 2017a, *Unification of the Eleven Boson Interactions based on 'Rotations of Interactions'* (Pingree Hill Publishing, Auburn, NH, 2017).

_____, 2017b, *The Origin of Fermions and Bosons, and Their Unification* (Pingree Hill Publishing, Auburn, NH, 2017).

_____, 2017c, *Megaverse: The Universe of Universes* (Pingree Hill Publishing, Auburn, NH, 2017).

_____, 2017d, *SuperSymmetry and the Unified SuperStandard Model* (Pingree Hill Publishing, Auburn, NH, 2017).

_____, 2017e, *From Qubits to the Unified SuperStandard Model with Embedded SuperStrings: A Derivation* (Pingree Hill Publishing, Auburn, NH, 2017).

_____, 2017f, *The Unified SuperStandard Model in Our Universe and the Megaverse: Quarks, ... ,* (Pingree Hill Publishing, Auburn, NH, 2017).

_____, 2018a, *The Unified SuperStandard Model and the Megaverse SECOND EDITION A Deeper Theory based on a New Particle Functional Space that Explicates Quantum Entanglement Spookiness (Volume 1)* (Pingree Hill Publishing, Auburn, NH, 2018).

_____, 2018b, *Cosmos Creation: The Unified SuperStandard Model, Volume 2, SECOND EDITION* (Pingree Hill Publishing, Auburn, NH, 2018).

_____, 2018c, *God Theory* (Pingree Hill Publishing, Auburn, NH, 2018).

_____, 2018d, *Immortal Eye: God Theory: Second Edition* (Pingree Hill Publishing, Auburn, NH, 2018).

_____, 2018e, *Unification of God Theory and Unified SuperStandard Model THIRD EDITION* (Pingree Hill Publishing, Auburn, NH, 2018).

_____, 2019a, *Calculation of: QED α = 1/137, and Other Coupling Constants of the Unified SuperStandard Theory* (Pingree Hill Publishing, Auburn, NH, 2019).

_____, 2019b, *Coupling Constants of the Unified SuperStandard Theory SECOND EDITION* (Pingree Hill Publishing, Auburn, NH, 2019).

_____, 2019c, *New Hybrid Quantum Big_Bang–Megaverse_Driven Universe with a Finite Big Bang and an Increasing Hubble Constant* (Pingree Hill Publishing, Auburn, NH, 2019).

_____, 2019d, *The Universe, The Electron and The Vacuum* (Pingree Hill Publishing, Auburn, NH, 2019).

_____, 2019e, *Quantum Big Bang – Quantum Vacuum Universes (Particles)* (Pingree Hill Publishing, Auburn, NH, 2019).

_____, 2019f, *The Exact QED Calculation of the Fine Structure Constant Implies ALL 4D Universes have the Same Physics/Life Prospects* (Pingree Hill Publishing, Auburn, NH, 2019).

_____, 2019g, *Unified SuperStandard Theory and the SuperUniverse Model: The Foundation of Science* (Pingree Hill Publishing, Auburn, NH, 2019).

_____, 2020a, *Quaternion Unified SuperStandard Theory (The QUeST) and Megaverse Octonion SuperStandard Theory (MOST)* (Pingree Hill Publishing, Auburn, NH, 2020).

_____, 2020b, *United Universes Quaternion Universe - Octonion Megaverse* (Pingree Hill Publishing, Auburn, NH, 2020).

_____, 2020c, *Unified SuperStandard Theories for Quaternion Universes & The Octonion Megaverse* (Pingree Hill Publishing, Auburn, NH, 2020).

_____, 2020d, *The Essence of Eternity: Quaternion & Octonion SuperStandard Theories* (Pingree Hill Publishing, Auburn, NH, 2020).

_____, 2020e, *The Essence of Eternity II* (Pingree Hill Publishing, Auburn, NH, 2020).

_____, 2020f, *A Very Conscious Universe* (Pingree Hill Publishing, Auburn, NH, 2020).

_____, 2020g, *Hypercomplex Universe* (Pingree Hill Publishing, Auburn, NH, 2020).

_____, 2020h, *Beneath the Quaternion Universe* (Pingree Hill Publishing, Auburn, NH, 2020).

_____, 2020i, *Why is the Universe Real? From Quaternion & Octonion to Real Coordinates* (Pingree Hill Publishing, Auburn, NH, 2020).

_____, 2020j, *The Origin of Universes: of Quaternion Unified SuperStandard Theory (QUeST); and of the Octonion Megaverse (UTMOST)* (Pingree Hill Publishing, Auburn, NH, 2020).

_____, 2020k, *The Seven Spaces of Creation: Octonion Cosmology* (Pingree Hill Publishing, Auburn, NH, 2020).

_____, 2020l, *From Octonion Cosmology to the Unified SuperStandard Theory of Particles* (Pingree Hill Publishing, Auburn, NH, 2020).

_____, 2021a, *Pioneering the Cosmos* (Pingree Hill Publishing, Auburn, NH, 2021).

_____, 2021b, *Pioneering the Cosmos II* (Pingree Hill Publishing, Auburn, NH, 2021).

_____, 2021c, *Beyond Octonion Cosmology* (Pingree Hill Publishing, Auburn, NH, 2021).

_____, 2021d, *Universes are Particles* (Pingree Hill Publishing, Auburn, NH, 2021).

_____, 2021e, *Octonion-like dna-based life, Universe expansion is decay, Emerging New Physics* (Pingree Hill Publishing, Auburn, NH, 2021).

_____, 2021f, *The Science of Creation New Quantum Field Theory of Spaces* (Pingree Hill Publishing, Auburn, NH, 2021).

_____, 2021g, *Quantum Space Theory With Application to Octonion Cosmology & Possibly To Fermionic Condensed Matter* (Pingree Hill Publishing, Auburn, NH, 2021).

108   **REFERENCES**

_____, 2021h, *21ˢᵗ Century Natural Philosophy of Octonion Cosmology , and Predestination, Fate, and Free Will* (Pingree Hill Publishing, Auburn, NH, 2021).

_____, 2021i, *Beyond Octonion Cosmology II : Origin of the Quantum; A New Generalized Field Theory (GiFT); A Proof of the Spectrum of Universes; Atoms in Higher Universes* (Pingree Hill Publishing, Auburn, NH, 2021).

_____, 2021j, *Integration of General Relativity and Quantum Theory: Octonion Cosmology, GiFT, Creation/Annihilation Spaces CASe, Reduction of Spaces to a Few Fermions and Symmetries in Fundamental Frames* (Pingree Hill Publishing, Auburn, NH, 2021).

Eddington, A. S., 1952, *The Mathematical Theory of Relativity* (Cambridge University Press, Cambridge, U.K., 1952).

Fant, Karl M., 2005, *Logically Determined Design: Clockless System Design With NULL Convention Logic* (John Wiley and Sons, Hoboken, NJ, 2005).

Feinberg, G. and Shapiro, R., 1980, *Life Beyond Earth: The Intelligent Earthlings Guide to Life in the Universe* (William Morrow and Company, New York, 1980).

Gelfand, I. M., Fomin, S. V., Silverman, R. A. (tr), 2000, *Calculus of Variations* (Dover Publications, Mineola, NY, 2000).

Giaquinta, M., Modica, G., Souchek, J., 1998, *Cartesian Coordinates in the Calculus of Variations* Volumes I and II (Springer-Verlag, New York, 1998).

Giaquinta, M., Hildebrandt, S., 1996, *Calculus of Variations* Volumes I and II (Springer-Verlag, New York, 1996).

Gradshteyn, I. S. and Ryzhik, I. M., 1965, *Table of Integrals, Series, and Products* (Academic Press, New York, 1965).

Heitler, W., 1954, *The Quantum Theory of Radiation* (Claendon Press, Oxford, UK, 1954).

Huang, Kerson, 1992, *Quarks, Leptons & Gauge Fields 2ⁿᵈ Edition* (World Scientific Publishing Company, Singapore, 1992).

Jost, J., Li-Jost, X., 1998, *Calculus of Variations* (Cambridge University Press, New York, 1998).

Kaku, Michio, 1993, *Quantum Field Theory*, (Oxford University Press, New York, 1993).

Kirk, G. S. and Raven, J. E., 1962, *The Presocratic Philosophers* (Cambridge University Press, New York, 1962).

Landau, L. D. and Lifshitz, E. M., 1987, *Fluid Mechanics 2ⁿᵈ Edition*, (Pergamon Press, Elmsford, NY, 1987).

Misner, C. W., Thorne, K. S., and Wheeler, J. A., 1973, *Gravitation* (W. H. Freeman, New York, 1973).

Rescher, N., 1967, *The Philosophy of Leibniz* (Prentice-Hall, Englewood Cliffs, NJ, 1967).

Rieffel, Eleanor and Polak, Wolfgang, 2014, *Quantum Computing* (MIT Press, Cambridge, MA, 2014).

Riesz, Frigyes and Sz.-Nagy, Béla, 1990, *Functional Analysis* (Dover Publications, New York, 1990).

Sagan, H., 1993, *Introduction to the Calculus of Variations* (Dover Publications, Mineola, NY, 1993).

Sakurai, J. J., 1964, *Invariance Principles and Elementary Particles* (Princeton University Press, Princeton, NJ, 1964).

Weinberg, S., 1972, *Gravitation and Cosmology* (John Wiley and Sons, New York, 1972).

Weinberg, S., 1995, *The Quantum Theory of Fields Volume I* (Cambridge University Press, New York, 1995).

INDEX

About the Author

Stephen Blaha is a well-known Physicist and Man of Letters with interests in Science, Society and civilization, the Arts, and Technology. He had an Alfred P. Sloan Foundation scholarship in college. He received his Ph.D. in Physics from Rockefeller University. He has served on the faculties of several major universities. He was also a Member of the Technical Staff at Bell Laboratories, a manager at the Boston Globe Newspaper, a Director at Wang Laboratories, and President of Blaha Software Inc. and of Janus Associates Inc. (NH).

Among other achievements he was a co-discoverer of the "r potential" for heavy quark binding developing the first (and still the only demonstrable) non-Aeolian gauge theory with an "r" potential; first suggested the existence of topological structures in superfluid He-3; first proposed Yang-Mills theories would appear in condensed matter phenomena with non-scalar order parameters; first developed a grammar-based formalism for quantum computers and applied it to elementary particle theories; first developed a new form of quantum field theory without divergences (thus solving a major 60 year old problem that enabled a unified theory of the Standard Model and Quantum Gravity without divergences to be developed); first developed a formulation of complex General Relativity based on analytic continuation from real space-time; first developed a generalized non-homogeneous Robertson-Walker metric that enabled a quantum theory of the Big Bang to be developed without singularities at t = 0; first generalized Cauchy's theorem and Gauss' theorem to complex, curved multi-dimensional spaces; received Honorable Mention in the Gravity Research Foundation Essay Competition in 1978; first developed a physically acceptable theory of faster-than-light particles; first derived a composition of extremums method in the Calculus of Variations; first quantitatively suggested that inflationary periods in the history of the universe were not needed; first proved Gödel's Theorem implies Nature must be quantum; provided a new alternative to the Higgs Mechanism, and Higgs particles, to generate masses; first showed how to resolve logical paradoxes including Gödel's Undecidability Theorem by developing Operator Logic and Quantum Operator Logic; first developed a quantitative harmonic oscillator-like model of the life cycle, and interactions, of civilizations; first showed how equations describing superorganisms also apply to civilizations. A recent book shows his theory applies successfully to the past 14 years of history and to *new* archaeological data on Andean and Mayan civilizations as well as Early Anatolian and Egyptian civilizations.

He first developed an axiomatic derivation of the form of The Standard Model from geometry – space-time properties – The Unified SuperStandard Model. It unifies all the known forces of Nature. It also has a Dark Matter sector that includes a Dark ElectroWeak sector with Dark doublets and Dark gauge interactions. It uses quantum coordinates to remove infinities that crop up in most

interacting quantum field theories and additionally to remove the infinities that appear in the Big Bang and generate inflationary growth of the universe. It shows gravity has a MOND-like form without sacrificing Newton's Laws. It relates the interactions of the MOND-like sector of gravity with the r-potential of Quark Confinement. The axioms of the theory lead to the question of their origin. We suggest in the preceding edition of this book it can be attributed to an entity with God-like properties. We explore these properties in "God Theory" and show they predict that the Cosmos exists forever although individual universes (or incarnations of our universe) "come and go." Several other important results emerge from God Theory such a functionally triune God. The Unified SuperStandard Theory has many other important parts described in the Current Edition of *The Unified SuperStandard Theory* and expanded in subsequent volumes.

Blaha has had a major impact on a succession of elementary particle theories: his Ph.D. thesis (1970), and papers, showed that quantum field theory calculations to all orders in ladder approximations could not give scaling deep inelastic electron-nucleon scattering. He later showed the eigenvalue equation for the fine structure constant α in Johnson-Baker-Willey QED had a zero at $\alpha = 1$ not $1/137$ by solving the Schwinger-Dyson equations to all orders in an approximation that agreed with exact results to 4th order in α thus ending interest in this theory. In 1979 at Prof. Ken Johnson's (MIT) suggestion he calculated the proton-neutron mass difference in the MIT bag model and found the result had the wrong sign reducing interest in the bag model. These results all appear in Physical Review papers. In the 2000's he repeatedly pointed out the shortcomings of SuperString theory and showed that The Standard Model's form could be derived from space-time geometry by an extension of Lorentz transformations to faster than light transformations. This deeper space-time basis greatly increases the possibility that it is part of THE fundamental theory. Recently, Blaha showed that the Weak interactions differed significantly from the Strong, electromagnetic and gravitation interactions in important respects while these interactions had similar features, and suggested that ElectroWeak theory, which is essentially a glued union of the Weak interactions and Electromagnetism, possibly modulo unknown Higgs particle features, be replaced by a unified theory of the other interactions combined with a stand-alone Weak interaction theory. Blaha also showed that, if Charmonium calculations are taken seriously, the Strong interaction coupling constant is only a factor of five larger than the electromagnetic coupling constant, and thus Strong interaction perturbation theory would make sense and yield physically meaningful results.

In graduate school (1965-71) he wrote substantial papers in elementary particles and group theory: The Inelastic E- P Structure Functions in a Gluon Model. Phys. Lett. B40:501-502,1972; Deep-Inelastic E-P Structure Functions In A Ladder Model With Spin 1/2 Nucleons, Phys.Rev. D3:510-523,1971; Continuum Contributions To The Pion Radius, Phys. Rev. 178:2167-2169,1969; Character Analysis of U(N) and SU(N), J. Math. Phys. 10, 2156 (1969); and The Calculation of the Irreducible Characters of the Symmetric Group in Terms of the

Compound Characters, (Published as Blaha's Lemma in D. E. Knuth's book: *The Art of Computer Programming Vols. 1 – 4*).

In the early 1980's Blaha was also a pioneer in the development of UNIX for financial, scientific and Internet applications: benchmarked UNIX versions showing that block size was critical for UNIX performance, developing financial modeling software, starting database benchmarking comparison studies, developing Internet-like UNIX networking (1982) and developing a hybrid shell programming technique (1982) that was a precursor to the PERL programming language. He was also the manager of the AT&T ten-year future products development database. His work helped lead to commercial UNIX on computers such as Sun Micros, IBM AIX minis, and Apple computers.

In the 1980's he pioneered the development of PC Desktop Publishing on laser printers and was nominated for three "Awards for Technical Excellence" in 1987 by PC Magazine for PC software products that he designed and developed.

Recently he has developed a theory of Megaverses – actual universes of which our universe is one – with quantum particle-like properties based on the Wheeler-DeWitt equation of Quantum Gravity. He has developed a theory of a baryonic force, which had been conjectured many years ago, and estimated the strength of the force based on discrepancies in measurements of the gravitational constant G. This force, operative in D-dimensional space, can be used to escape from our universe in "uniships" which are the equivalent of the faster-than-light starships proposed in the author's earlier books. Thus travel to other universes, as well as to other stars is possible.

Blaha also considered the complexified Wheeler-DeWitt equation and showed that its limitation to real-valued coordinates and metrics generated a Cosmological Constant in the Einstein equations.

Blaha has been developing hypercomplex cosmologies from January, 2020 to the present. He has published his work in a series of books.

The author has also recently written a series of books on the serious problems of the United States and their solution as well as a book on the decline of Mankind that will follow from current social and genetic trends in Mankind.

In the past twenty years Dr. Blaha has written over 80 books on a wide range of topics. Some recent major works are: *From Asynchronous Logic to The Standard Model to Superflight to the Stars, All the Universe!, SuperCivilizations: Civilizations as Superorganisms, America's Future: an Islamic Surge, ISIS, al Qaeda, World Epidemics, Ukraine, Russia-China Pact, US Leadership Crisis, The Rises and Falls of Man – Destiny – 3000 AD: New Support for a Superorganism MACRO-THEORY of CIVILIZATIONS From CURRENT WORLD TRENDS and NEW Peruvian, Pre-Mayan, Mayan, Anatolian, and Early Egyptian Data, with a Projection to 3000 AD*, and *Mankind in Decline: Genetic Disasters, Human-Animal Hybrids, Overpopulation, Pollution, Global Warming, Food and Water Shortages, Desertification, Poverty, Rising Violence, Genocide, Epidemics, Wars, Leadership Failure*.

He has taught approximately 4,000 students in undergraduate, graduate, and postgraduate corporate education courses primarily in major universities, and large companies and government agencies.

Recently he developed a quantum theory, The Unified SuperStandard Theory (UST), which describes elementary particles in detail without the difficulties of conventional quantum field theory. He found that the internal symmetries of this theory could be exactly derived from an octonion theory called QUeST. He further found that another octonion theory (UTMOST) describes the Megaverse. It can hold QUeST universes such as our own universe. It has an internal symmetry structure which is a superset of the QUeST internal symmetries.

CPSIA information can be obtained
at www.ICGtesting.com
Printed in the USA
BVHW010843060222
627286BV00028B/83

9 781737 264095